Organizational Sustainability and Risk Management

This new edition is completely revamped and reorganized to reflect the change in standards and regulations and to include all new topics related to organizational sustainability and risk management. The role that the Sustainable Development Goals (SDGs) play within the realm of organizational sustainability is one of many new topics.

Organizational Sustainability and Risk Management: A Practical Step-by-Step Guide, Second Edition will continue to remind all stakeholders how organizations work through a measurement transformation that affects everything they do including following the International Organization for Standardization's (ISO) Guide for Sustainability and climate change. The book is enriched with a discussion on life cycle thinking that has been introduced in the ISO high-level structure. Discussions on a fundamental change in how organizations approach sustainability and how we view organizational sustainability are covered. This book offers a platform for managing all activities, products, and services tailored to the needs of the organization and presents how important environmental, social, and governance (ESG) standards are to determine the potential for increased financial growth of organizations that have implemented organizational sustainability.

The book is for professionals and can be used in continuing education sustainability courses as well as company-provided short courses where the new regulations for sustainability and ESG reporting are addressed.

Organizational Sustainability and Risk Management

A Practical Step-by-Step Guide

Second Edition

Denice Viktoria Staaf and Robert B. Pojasek

CRC Press
Taylor & Francis Group
Boca Raton London New York

CRC Press is an imprint of the
Taylor & Francis Group, an **informa** business

Boca Raton and London

Second edition published 2024

by CRC Press
6000 Broken Sound Parkway NW, Suite 300, Boca Raton, FL 33487-2742

and by CRC Press
4 Park Square, Milton Park, Abingdon, Oxon, OX14 4RN

CRC Press is an imprint of Taylor & Francis Group, LLC

© 2024 Denice Viktoria Staaf, Robert B. Pojasek

First edition published by CRC Press 2017

Library of Congress Cataloging-in-Publication Data
Names: Staaf, Denice Viktoria, author. | Pojasek, Robert B., author.
Title: Organizational sustainability and risk management : a practical
step-by-step guide / Denice Viktoria Staaf, Robert B. Pojasek.
Other titles: Organizational risk management and sustainability.
Description: Second Edition. | Boca Raton : CRC Press, 2024. | Revised
edition of Organizational risk management and sustainability, [2017] |
Includes bibliographical references and index.
Identifiers: LCCN 2023001630 (print) | LCCN 2023001631 (ebook) | ISBN
9781032185545 (hardback) | ISBN 9781032185576 (paperback) | ISBN
9781003255116 (ebook)
Subjects: LCSH: Risk management. | Management. | Sustainability.
Classification: LCC HD61 .P635 2024 (print) | LCC HD61 (ebook) | DDC
658.15/5--dc23/eng/20230119
LC record available at https://lccn.loc.gov/2023001630
LC ebook record available at https://lccn.loc.gov/2023001631

ISBN: 978-1-032-18554-5 (hbk)
ISBN: 978-1-032-18557-6 (pbk)
ISBN: 978-1-003-25511-6 (ebk)

DOI: 10.1201/9781003255116

Typeset in Times
by Deanta Global Publishing Services, Chennai, India

Dedication

For Dr. Robert B. Pojasek, the sustainability legend who taught thousands of professionals globally during his 20 years teaching online at Harvard University. Dr. P, you are an incredible teacher, mentor, and friend. May your wisdom and passion for sustainability be infused into all who read this book in the ways you gifted it to me. I cannot thank you enough for your patience, kindness, and friendship. I am grateful.

Contents

Preface

This is the second edition of the book, released in February 2017. It was just in time to be used in the "Organizational Sustainability" course at Harvard University's School of Continuing Education. The concept was that the term "sustainability" needs to have a qualifier when used in conversation and writing. We began influencing this feeling with "organizational" as the qualifier. Other books and industry programs still focus on "corporate" sustainability or use the company name as the qualifier. During this period, the term "CSR," or Corporate Social Responsibility, was used less frequently.

In the past three years, there has been a greater focus on "climate change" and how larger companies and their supply chains influence it. Stakeholders look closely at the products and services associated with the organization that have "greenhouse gases" associated with the corporation. The use and loss of greenhouse gas emissions need to be controlled by changes instituted through the management, the employees, the suppliers, and the customers of companies with their organizational sustainability program. The internal and external practices of organizations should have a strong social foundation.

Organizational sustainability is part of how people work every day! It is also the focus that citizens and employees want organizations to have. As we started working on the book, the International Financial Reporting Standards Foundation (IFRS Foundation) led the formation of the International Sustainability Standards Board (ISSB) to extend organizational sustainability to involve the "Capital Markets better." We were hoping to include more than just a mention in this volume. The focus on financial issues (e.g., climate change) will soon be a measurable action captured by all publicly traded companies and their responsibility to charge the costs of ALL operations associated with the company's products and services. These efforts will just be starting as this book is published. We have included some basic information on this next "big wave" of organizational sustainability to help you understand what will happen between this edition and the need for a third edition in 2025. However, the ISSB will need to pay particular attention to this version not to lose the social aspects of organizational sustainability as they begin quantifying the financial aspects of "organizational sustainability."

Now that the second edition is in print, it would be worthwhile to finish the work of moving from "corporate sustainability" to "organizational sustainability." This book provides the foundation and language to develop the processes needed in your organization. We are now beginning to justify the move to sustainable finance that will offer awards to those corporate leaders trying to bring this all together as described above.

Robert B. Pojasek, Ph.D.
December 2022

Authors

Denice Viktoria Staaf

Denice Viktoria Staaf found her way to sustainability through the green building movement. She has been an Accredited Professional under the Leadership in Energy and Environmental Design (LEED) Standard for over 20 years. She is passionate about sustainability for manufacturers and supply chains under the overall organizational sustainability umbrella. She believes that the systematic approach taught by Dr. Pojasek, outlined in this book, is the way forward for organizations to contribute to sustainable development positively. She graduated from Harvard University, earning a master's degree in Environmental Management & Sustainability with a concentration in the Built Environment.

Denice has recently focused on circularity as a deeper dive into organizational sustainability using the process approach to achieve results. She is a member of the Circular Transition Indicators (CTI) Framework Implementation Partners by the World Business Council for Sustainable Development (WBCSD) and recently joined the International Organization for Standardization (ISO) Technical Committee for the Circular Economy (TC 323).

Denice founded the consulting firm Labeling Sustainability in 2016 after helping companies of her friends with sustainability for fun; she loves organizational sustainability and being of service to manufacturers. Today, Labeling Sustainability is a global consulting firm and Type III Environmental Product Declaration Program Operator assisting international organizations with their sustainability documentation needs. Although the organization has grown from the days she did LCAs for her friends, Denice Viktoria Staaf still thinks sustainability is fun.

Robert B. Pojasek, PhD

Dr. Pojasek has been active in the reformatting of the corporate sustainability reporting by using the organizational structure that is presented in the first volume of this work, i.e., "organizational sustainability." To make this happen, I have worked closely with Denice Viktoria Staaf, who has introduced herself above.

Dr. Pojasek has been interested in the organizational practice of organizational sustainability since the lead-up to the 1992 Earth Summit held in Rio de Janeiro. The United Nations was successful in convincing large companies to get involved in the concept of sustainability as the United States and Europe began introducing legislation on providing communities with clean air and water. Prior to this event, Dr. Pojasek had spent 12 years developing pollution prevention programs in the companies. The interest in "sustainability" and the experience of working with facilities to eliminate or reduce the "waste" in their operations was just beginning. However, consulting firms and university research were slow to develop. Consulting firms were focused on the permitting of processes, not on the elimination of their use.

Dr. Pojasek began teaching pollution prevention at Tufts University, with an eye on developing what people defined as "sustainability." He worked to build interest in this topic by publishing over 100 journal articles on the topic and inviting industry managers to take the courses that he was teaching at Tufts. In 200, he worked with several professors at the Harvard Extension School to develop a "master's degree program" that would require courses that he wanted to teach. The program began in 2002. Dr. Pojasek developed a course in "Organizational Sustainability" and taught that course for 20 years. He is still working hard to have more practical courses, books, and experiences with organizational sustainability that will soon be supplemented with "sustainable finance." That is the new frontier of this developing story. It is a little soon to be introducing volume 3 of this series. Maybe the first two volumes (volume 1 is in paperback) will convince you to work with corporations to make the changes to adopt the new formats for organizational and financial sustainability.

1 The Path to Organizational Sustainability

ABSTRACT

Sustainability was first proposed as the goal of sustainable development. Sustainability has three dimensions—environmental, social, and economical. These dimensions are mutually reinforcing and interdependent. It is a misunderstanding to limit sustainability to a single dimension, such as climate change, unsustainable resource depletion, or biodiversity. Sustainability is relevant to all levels of human activity, from the global level to the national, regional, community, organizational, and individual levels. Since organizations are the basic building blocks of society, it is important to see how this perspective differs from the other definitions.

IT STARTED WITH THE BRUNDTLAND REPORT

In the 1980s, development and environmental improvement seemed to be failing. Since the first Earth Day in 1970, the public interest was growing in each topic. The "Superfund" legislation in the United States began the slow process of remediating "hazardous waste dumpsites." In addition, the countries below the Equator were experiencing a poverty crisis associated with many developmental and environmental problems that had the attention of citizens' groups worldwide. Neither the UN Development Program (UNDP) nor the UN Environmental Program (UNEP) could effectively address these problems. This situation reached the UN General Assembly, which in 1983 adopted a resolution to establish a commission to seek ways to find a path forward. Thus, the World Commission on Environment and Development (WCED) was formed with the support of the Norwegian Environment Minister, Gro Harlem Brundtland, as the Chair.

WCED is best known for preparing the 1987 publication that explored the interconnections between social equity, economic growth, and a wide range of cases involving the causes of environmental degradation. This information was released in a book entitled *Our Common Future*.[1] This book is still available on the Internet.

The Brundtland Commission's definition of sustainable development[2] is:

> Development that meets the environmental, social and economic needs of the present without compromising the ability of future generations to meet their own needs.[2]

As more discussions took place, sustainable development looked to understanding and achieving a balance between environmental, social, and economic systems.

DOI: 10.1201/9781003255116-1

This was considered essential for making progress toward achieving sustainability. However, it was the second term that would cause the most challenging acceptance problem:

> State of the global system, including environmental, social, and economic aspects, in which the needs of the present are met without compromising the ability of future generations to meet their own needs. [2]

This term was more difficult to absorb because it defined "global sustainability" instead of organizational sustainability. Given the intergenerational nature of global sustainability and the constant changes in the environmental, societal (e.g., population growth), and economic subsystems, global sustainability cannot be described purely in terms of a single fixed endpoint. From this perspective, sustainability is a characteristic of the planet, not any activity or organization. Yet, sustainable development addresses the activities of organizations (i.e., businesses, communities, governments, and unilateral organizations) in a manner that contributes to sustainability. Such development is needed to meet the needs of both present and future generations. While this is essential, we are left without a definition of "organizational sustainability." No one was thinking about it at that time.

THE EARTH SUMMIT IN 1992

One of the messages of the Brundtland Report involved launching a plan for an international conference to discuss the content in the report. This conference occurred at the Rio de Janeiro Earth Summit in 1992. One hundred seventy-eight countries were represented, and more than 100 country leaders participated. At the end of the conference, everyone left with an understanding of sustainable development. Global sustainability was more difficult to absorb because it defined the entire globe instead of the work that could be controlled by industry and government. Given the intergenerational nature of global sustainability and the constant changes in environmental, societal, and economic subsystems, global sustainability cannot be described purely in terms of a single fixed endpoint. From this perspective, global sustainability is a characteristic of the planet, not any activity or organization. Yet, sustainable development addresses the activities of the organization (i.e., business, communities, governments, and unilateral organizations) in a manner that contributes to sustainability over time. Such development is needed to meet the needs of both present and future generations. It is therefore essential to global sustainability.

Knowing early that industry needed to be involved in these activities, Maurice Strong[3] wanted to ensure that industry was represented. He asked Stephan Schmidheiny (Swiss Industries) to serve as the event's principal adviser for business and industry. Schmidheiny conscripted a team of 48 leaders of major companies to create a group with two functions: to spread the sustainable development message among businesses and produce business input for the deliberations at the Earth Summit. This group was named the Business Council for Sustainable Development

(BCSD). The International Chamber of Commerce also created some input for the Earth Summit.

The members of BCSD understood that if all organizations were able to address sustainable development in their operations, global sustainability would be the result. However, they also understood that global sustainability meant that sustainable development needed to be contributed to by all organizations working with "sustainability" as an objective. Therefore, they required a means of finding a practice that would contribute to sustainable development.

ECO-EFFICIENCY

The BCSD started searching for a mission to pave the way for sustainable development as part of the business. It was clear what was in store for the governments and communities. What about business! They seemed to be left out of the overall planning of the meeting, with nothing to be the focus of their planning. At their first meeting, about a year before the Earth Summit, a couple of members came up with the term "Eco-efficiency." That would be as difficult to act as sustainable development and global sustainability. With this and other ideas, Stephan Schmidheiny began to determine what could be added to the meeting's signature document—"Agenda 21"[4]—on issues including clean production techniques, energy use, pricing instruments, capital markets, and managing agriculture and forestry. Some BCSD members took on some of the writing so that the industry could gather some praise to move forward with what they could contribute to the Earth Summit. It was going to be challenging to develop the consensus that was needed.

Schmidheiny's mandate from Maurice Strong was to bring a business voice to Rio and to spread the concept of sustainable development to the world's business leaders and companies. Over 50 conferences, symposia, and issue workshops took place in 20 countries to accomplish the second goal during the Earth Summit. It was challenging to get any consensus from all 50 members of BCSD in such a short time. The differences over many of the items were vast. However, many felt their participation in BCSD "changed their life." No one was thinking about sustainability before. Schmidheiny wrote *Changing Course*[5] to ensure that the business involvement was recognized at the Earth Summit. This book was peer-reviewed at MIT and was distributed to the meeting participants about a month before the meeting.

The Earth Summit's goals were ambitious and focused on providing help to governments to rethink economic development and find ways to halt pollution and the destruction of natural resources. It was a fantastic event and the largest gathering of world leaders in history, with 117 heads of state and representatives of 178 nations. Attendees agreed to try to stabilize greenhouse gas, lower the production of toxic components and wastes, switch from fossil fuels to alternative energy sources, rely more on public transportation, and give more attention to the growing water scarcity. "Agenda 21" provided a comprehensive plan of action to be taken globally, nationally, and locally by organizations of the United Nations system, governments, and other significant groups, a wide-ranging blueprint for action to achieve sustainable development worldwide. However, while most marveled with the document, it was widely considered weakened by compromise and a lack of priorities.

However, even though the industry did not develop a clear role to contribute to a more sustainable way of doing business, they got the conversation going. Unfortunately, because the industry was still considering what it needed to do while maintaining the vitality of its businesses, there was no universal and accepted guide to sustainability for the industry as a whole.

INDUSTRY TRIES TO DEFINE ITS VERSION OF SUSTAINABILITY

Over the next few years, BCSD tried to define its vision of what would become known as "corporate sustainability." Many people involved in the Earth Summit dropped out of the group. BCSD struggled to find a clear target to guide their efforts. They encouraged the International Organization for Standardization (ISO) to complete their work on a standard for environmental management. They changed the leadership focus from the CEO commitment to a corporate commitment. Several groups working on the issues merged in 1995 to create the World Business Council for Sustainable Development (WBCSD).

At one of the meetings, someone from Shell[6] remarked that if the WBCSD were serious about understanding the future of sustainability and business, it should do scenario planning. They discussed many ways in which an issue might play out in the future using this tool. Shell asked WBCSD what role they wish to play in helping shape that future. After the meeting, a few members convened to discuss scenario planning and got the go-ahead to approach Shell. Shell agreed that WBCSD would pay US$ 1 million to fund the project. Over the following year, they raised US$ 750,000, and Shell agreed to cover the remaining US$ 250,000.

Three scenarios were created and presented at a WBCSD meeting. The response was "electric!" At the urging of Shell, the WBCSD scenarios were taken on tour for a year and presented to groups of companies, intergovernmental groups, and the World Bank. The reactions were very positive. It was argued that sustainability could be what quality was 15 years prior. The scenarios elevated the perception of the WBCSD on the global stage from being the business voice to being a thoughtful commentator on the complex roles of all the players in the sustainable development area. They became the thought leaders on this topic worldwide when companies published their environmental and health and safety annual reports and the corporation's annual report. The context of this activity significantly informed the work of WBCSD and its members.

CORPORATE SOCIAL RESPONSIBILITY

When the business voice was first recognized at the Earth Summit, the framework of issues around sustainability was primarily restricted to the "environment" and the actual cost of pollution to the economy as it occurred in the mid-1990s. This was modified by the bottom-line gains from increased energy and resource efficiency. No universally accepted definition of CSR existed. People often say that it is about what a business puts back in return for the benefits it receives from society. Others will say that CSR is about a company's interaction with the legal and social obligations of the societies in which it operates and about how it accounts for these obligations.

The Global Reporting Initiative (GRI)[7] was founded in 1997, soon after the public outcry over the environmental damage of the *Exxon Valdez* oil spill. GRI was an ambitious attempt to set voluntary standards for corporate responsibility reporting and an excellent example of a successful partnership between business, a UN organization (UN Environmental Program), and NGOs. It launched its first voluntary reporting standard in 2000.

At the same time, a series of international corporate scandals broadened that focus to include social concerns around sweatshops and supply chains. As a result, the social component of sustainability found its place in the move to improve the corporate governance processes. As a result, the industry responded with "Corporate Social Responsibility" (CSR), which describes activities ranging from corporate philanthropy to positive moves to improve and sustain labor practices.

Navigating this area meant grappling with the wants and needs of an ever-widening range of stakeholders, including customers, partners, suppliers, the community, the environment, and future generations. In 1999, the WBCSD issued its first report on CSR. It was entitled "Corporate Social Responsibility – Meeting Changing Expectations."[8] The report addressed human rights, employee rights, environmental protection, community involvement, and supplier relations. The WBCSD refused to officially define CSR by providing "insight" on how they viewed the topic early in their journey.

> Corporate social responsibility is the commitment of business to contribute to sustainable economic development, working with employees, their families, the local community, and society to improve their quality of life.

The year 1999 also witnessed the launch of the UN Global Compact.[9] A consortium of worldwide businesses and other interests committed to improving CSR performance. Of the initial 38 compact signatories, 15 were WBCSD members.

Sustainability/CSR annual reports became a way for companies to communicate with their stakeholders about the voluntary program to address the growing body of information described above. Attached to the report would be a list of voluntary sustainability items that met the materiality requirements of their program. Most of the lists of sustainability/CSR reporting items had their materiality requirements.

CORPORATE SUSTAINABILITY

Many corporations have improved their EHS reporting system and found it helpful in developing engagement with the organization's internal and external stakeholders. Many researchers became involved in the general practice of corporate sustainability. They wrote comprehensive books on corporate sustainability; one example is *The Sustainability Handbook: The Complete Management Guide to Achieving Social, Economic, and Environmental Responsibility*.[10]. This book and similar ones are the beginning of the study of corporate sustainability.

ESG SUSTAINABILITY

The term "ESG sustainability" was first used in 2005. In January 2004, UN Secretary-General Kofi Annan sent letters to approximately 50 CEOs of major financial institutions to invite them to participate in a one-year joint initiative. They would play a significant role in searching for ways to integrate "ESG" into capital markets rather than solely in company sustainability reports. The result of this effort[11] was the "Who Cares Wins" report. This document made the case that embedding environmental, social, and governance factors in capital markets makes good business sense and leads to more sustainable markets.

Around the same time, UNEP/Fi[12] produced the "Fresh Field Report." Together with the "Who Cares Wins" report, these two efforts formed the springboard for the launch of the "Principles for Responsible Investment (PRI) at the New York Stock Exchange in 2006 and the launch of the Sustainable Stock Exchange Initiative in 2007[13] and the Sustainable Stock Exchange Initiative in 2007.

Today the PRI's role is to advance the integration of ESG into analysis and decision-making through thought leadership and the creation of tools, guidance, and engagements. The SSEI supported by the Geneva-based UNCTAD has grown over the years, with many exchanges now mandating ESG disclosure for listed companies or providing guidance on reporting on ESG issues. Despite the success of this reporting group placing them into the mainstream, the risk of ESG investing could not become popular until about 2018, when over 80% of the world's largest corporations began using it. It is a replacement for corporate sustainability, and a company is rated by one of the significant ESG rating firms.

ORGANIZATIONAL SUSTAINABILITY

This form of sustainability is taken from the perspective of an organization.[14]

"Sustainability is the capability of an organization to manage its responsibilities for environmental stewardship, social well-being transparently, and economic shared values over the long term, the whole being held accountable to its stakeholders." This definition is actionable with any organization at the community level and is not a euphemism, slogan, the color green, or a specific initiative.

To reinforce these responsibilities and make them more specific to daily activities, an organization can create a "code of conduct" that outlines expectations for how responsibility will be embedded into what employees do daily. Organizations may also specify responsibility as part of their core values to ensure that "acting responsibly" is part of their culture. These codes of conduct typically state that the organization should adhere to the following practices.

- Be accountable for its impact on the environment, society, and the economy
- Be transparent in its decisions and activities that impact its responsibilities
- Behave ethically
- Respect, consider, and respond to the interests of its stakeholders
- Accept that respect for the rule of law is mandatory

Each organization should have its list of significant sustainability responsibilities. While they are often divided by categories, please remember that each responsibility is interconnected with the others.

To be effective, the code of conduct should consider several essential items:

- Maintaining evidence of compliance with relevant local, regional, and federal laws
- Understanding the consequences of noncompliance with the laws and regulations
- Effectively managing the elements of the code of conduct
- Maintaining the integrity and reputation of the organization
- Adhering to aspirational values in line with organizational sustainability
- Dealing with conflicts of interest and confidentiality
- Being responsible for engaging with external stakeholders
- Maintaining nondiscriminatory practices
- Paying attention to how members or employees are treated
- Conditions of membership or employment
- Using accounting for organizational resources
- Conditions within the organization are safe and hygienic
- Paying attention to occupational health and safety
- Acting as a steward of the environment and a good community citizen

An organization's commitment to the code of conduct should include the benefits and importance of having such a code or honoring a supplier's code of conduct. In addition, the code needs to be an integral component of the framework developed to manage risk and organizational sustainability within the organization. For an organization to practice environmental stewardship, its activities, processes, decisions, products, and services should strive to have no negative consequences to the environment. Environmental responsibility involves the following:

- Having effective processes for all operations associated with products or services and a system of management to provide management oversight of the processes and operations (e.g., environmental, assets, energy management, business continuity, social responsibility, governance, health & safety, and sustainable development)
- Enhancing the productivity of natural resources—use only what is needed, use it efficiently within the focus on reducing or eliminating waste, and be aware of issues with the management of waste that is not eliminated, including the reuse of products at their end of life
- Being mindful of energy use and climate change and not simply switching to renewable energy, which has significant Scope 3 greenhouse gas emissions or large amounts of embodied energy in the fabrication, installation, operation, and maintenance of the technology
- Addressing the stewardship of the natural habitat and its biodiversity in the neighborhood, the community, and areas affected by the value chain partners.

In the case of social well-being, the organization should seek to avoid negative consequences to society, with particular emphasis on its members, employees, and other people directly impacted by its activities and processes. Decisions, products, and services. All its suppliers should be contractually obligated to follow their supplier code of conduct or have management systems in place to facilitate the progress to sustainable development.

Social well-being includes the following:

- Respect for human rights by having the organization exercise due diligence in determining where human rights issues may arise within its value chain
- Responsibility for its labor practices, both in its operations and where it has a sphere of influence in the value chain
- A management system (e.g., health and safety; governance, risk, compliance, social responsibility, and risk management) is in place to facilitate social well-being
- Adopting fair operating practices to deal ethically with other organizations, including preventing corruption, responsible participation in the political process, respecting property rights, and promoting responsibility in its sphere of influence
- Involvement in community development and participating in local education and culture, public health, illiteracy, social investment, and quality of life

Each organization should use its control or sphere of influence in partnership with other local organizations and its value chain partners to address economic issues and their interrelationships with the other two sustainability responsibilities:

- Employment in the community
- Poverty and similar needs
- Local business climate
- Income levels
- Economic performance and community development
- Use of technology and innovation
- Value and supply chain prosperity
- Maintenance of the social license to operate
- Working with other local organizations to promote the value of organizational sustainability

Here are items that could be included in an organization's list of responsibilities:

- Continually improve the resource productivity of the operations
- Eliminate wastes of all kinds
- Pay attention to the prevention side of the activity rather than using recycling or controls
- Manage energy to respect the need for climate change mitigation

- Protect natural habitats and biodiversity
- Consider environmental impacts in areas under the organization's control and within its sphere of influence
- Protect human rights with an evaluation of the entire value chain
- Ensure fair operating practices
- Assess labor practices, including health and safety
- Evaluate consumer issues associated with products and services
- Optimize community involvement
- Consider social impacts and license to operate in areas under the organization's control and within its sphere of influence
- Contribute to the community's development
- Look for opportunities to share value within the community
- Consider community shared value impacts in areas under the organization's control and sphere of influence

Attention to these items should help an organization operate responsibly

OPERATING RESPONSIBLY

Operating responsibly is at the core of an organization's sustainability program. As in the case of risk management, the three responsibilities must be integrated and embedded in the organization's activities, processes, decisions, services, and products. Responsibility is seen as a balanced approach for organizations to address environmental, social, and economic issues in a way that aims to benefit people, communities, and society. Organizations are responsible for the consequences of their activities and decisions through transparent and ethical behaviors. The responsibility extends to the customers, neighborhood, community, society, and environment. Exercising an organization's responsibility involves many aspects of its operations:

- Contribution to sustainable development, including health and welfare of the community and society
- Active engagement with stakeholders to determine their interests in the organization and its products and services
- Operating in a manner that complies with applicable laws and is consistent with the international norms of behavior
- Integrating sustainability throughout the organization and practicing it in relationships within its control or sphere of influence

Several specific relationships guide the responsibilities of an organization. They include the relationship between the organization and society, between the organization and its stakeholders, and between the stakeholders and society. All these relationships affect the operation of organizations at the community level.

First, all organizations should have a relationship with the community. Organizations must understand how processes, decisions, products, and services affect the community. Organizations often support the community as an excellent place for their employees or members to live. Usually, the suppliers have operations in the community as well.

Second, organizations have a relationship with their stakeholders—both inside and external to the organization. Just as each organization engages with the internal stakeholders, it is essential to extend the engagement beyond the customer to other external stakeholders. This dialogue with stakeholders should be face-to-face, interactive, and over long periods. The interests of the stakeholders need to be understood and acknowledged. Some form of mediation should be considered if there are issues with stakeholders. This will enable engagement to dominate the agenda with all stakeholders.

Third, the organization's stakeholders have a relationship with the larger community (state, province, regional, and federal). Since stakeholders can be associated with diverse groups, it is possible that some of their interests are not consistent with the expectations of the community at large. Therefore, stakeholder interests must be carefully balanced across a broad spectrum of interests and may conflict with their own.

Organizations must understand how these relationships can complicate their ability to maintain their social license to operate. To some degree, local organizations have always conducted their activities with particular awareness of their relationship with the community. However, with the range of communication methods available today, it is even more critical that they pay specific attention to these relationships. The organization must responsibly decide how it will embed sustainability into its operations, rather than focusing on "initiatives" that compete with its core day-to-day operational activity.

EMBEDDING SUSTAINABILITY

As with risk management, organizational sustainability should be part of what every member or employee does daily. At the parent organization level, claims are often made that sustainability is embedded within the entire organization's structure and functions. The reality is that very few parent organizations have fully embedded or integrated sustainability into the way they operate day in and day out. Sustainability is frequently operated as a separate program with objectives not aligned with the organization's strategic objectives. Many of these objectives are designed to appease outside interests and not for the point of operating in a stewardship mode, seeking to prevent creating environmental, social, and economic problems.

There are two different forms of embedding sustainability. The first involves making sustainability and the responsibilities associated with sustainability part of the work instructions and operational controls of everyone in the organization. Sustainability would be part of what they do and not practice solely as a separate activity (e.g., green team initiatives). The second way of embedding sustainability is to make its considerations part of every decision at all organizational levels. In either case, there needs to be a close connection between sustainability and the organization's strategic objectives. It is also essential for there to be shared value between the stakeholders and the organizations. This is more complicated at the parent organization scale than at the organizational level. The strategy comes from the mission statement in terms of the strategic objectives. These objectives are cascaded down to the lower levels of the organization. Workers have goals and an action plan to achieve them using the guidance and structure of the strategic objectives covering their work. The realization of the goals at each level of the organization can be compared with the objectives to see if value is created over and above meeting the objectives. Sustainability would need to be included within the strategic objectives and the focus of the action plans associated with every worker's goals.

All organization's members or workers must understand its strategic objectives. These objectives need to be transparent to both internal and external stakeholders. When goals are established, many of the responsibilities listed above can be incorporated as potential means for creating value and ensuring the effectiveness of the stewardship approach. The effects of uncertainty must be addressed if the organization wants to meet its strategic objectives every year. Organizational sustainability has many moving parts, but they are much easier to control at the organizational level. The rest of this book will show how this is accomplished.

ESSENTIAL QUESTIONS FOR ORGANIZATIONAL SUSTAINABILITY

1. Why did it take so many different attempts to move from the Earth Summit (1992) to 2012 to provide the name of "organizational sustainability" to the practice we are familiar with today?
2. How does an organization's detailed "code of conduct" help it embed the day-to-day efforts to practice the three dimensions of the practice: environmental, social, and economic?
3. How does an organization select activities in the three elements of environmental stewardship, social well-being, and economic shared value that help it succeed in its attempt to attain organizational sustainability?
4. What are the ways in which an organization can operate responsibly using this kind of sustainability program?

REFERENCES

1. G. Brundtland, *Our Common Future: Report of the World Commission on Environment and Development*, N-Dokument A/42/427, 1987.
2. Geneva, ISO (International Organization for Standardization), "Guidelines for addressing sustainability in standards," Guide 82, Geneva, 2014.
3. M. Strong, "Chapter 28 of agenda 21, UN, New York," 1992. [Online]. Available: Www .mauricestrong.net. from https: //www.mauricestrong.net/ United Nations.
4. B. Rio de Janeiro, "Statement of forest principles: The final text of agree," Agenda 21: programme of action for sustainable development, 1993. [Online]. Available: https://sus tainabledevelopment.un.org/outcomedocuments/agenda21.
5. Geneva, "World business council for sustainable development WBCSD," *Corporate Social Responsibility: Making Good Business Sense*, Geneva: WBCSD, 2000.
6. "Shell scenarios," *Shell Global* [Online]. Available: https://www.shell.com/energy-and -innovation/the-energy-future/scenarios.html.

7. "Continuous improvement," *GRI - Standards*, 1997. [Online]. Available: https://www.globalreporting.org/standards.

8. World Business Council for Sustainable Development, "Corporate Social Responsibility: Meeting Changing Expectations," WBCSD Publications, 1999. [Online]. . Available: http://www.wbcsd.org/pages/edocument/edocumentdetails.aspx?id=82&nosearchcontextkey=true

9. "UN global compact," [Online]. Available: https://www.unglobalcompact.org/.

10. W. R. Blackburn, *The Sustainability Handbook: The Complete Management Guide to Achieving Social, Economic and Environmental Responsibility*, London: Earthscan, 2012.

11. "Who cares wins connecting financial markets to a changing world," [Online]. Available: https://d306pr3pise04h.cloudfront.net/docs/issues_doc%2FFinancial_markets%2Fwho_cares_who_wins.pdf.

12. F. B. Deringer, "A legal framework for the integration of environmental, social and governance issues into institutional investment," 2005. [Online]. Available: https://www.unepfi.org/fileadmin/documents/freshfields_legal_resp_20051123.pdf.

13. "Sustainable stock exchanges," [Online]. Available: /https://sseinitiative.org/.

14. "Organizational sustainability.action inclusion," July 16 2015. [Online]. Available: https://actioninclusion.org/leadership-diversity-change/what-is-organizational-sustainability/.

15. Brundtland, G.H. "*Report of the World Commission on Environment and Development*," 1987. [Online]. Available: https://digitallibrary.un.org/record/139811?ln=en

16. Online. Available: http://wbcsdservers.org/wbcsdpublications/cd_files/datas/wbcsd/corporate/pdf/CatalyzingChange-A%20ShortHistoryOfTheWBCSD.pdf.

2 Organizational Risk Management

ABSTRACT

There are many definitions of risk; most focus on hazards, harm, and harmful events. Only one of these definitions is presented from the perspective of an organization. This definition covers what might happen and how it will affect an organization's ability to meet its objectives in an uncertain world. No longer will risk be confined to harmful events. Instead, risk and its consequences could be positive (upside of risk) or negative (downside of risk). Risk is always focused on the organization's objectives. Opportunities and threats represent the effects of uncertainty.

UNDERSTANDING ORGANIZATIONAL RISK

Risk is the effect of uncertainty on objectives. This definition shifts the emphasis from an event (something happens) to the effects of uncertainty.[1–6] Risk is associated with strategic objectives. The effects of uncertainty can help the organization (opportunities) or hinder the organization (threats) from meeting its objectives. In the absence of uncertainty, risk comes from the execution of effective processes, efficient operations, and efficacious strategy.

After setting the strategic objectives, organizations must address the internal and external factors that can generate uncertainty. This is accomplished when the organization prepares a scan of the external and internal operating environments. These operating environments are known as the context of the organization. In the world of financial risk, organizations seek to avoid, control, or transfer to others (e.g., purchase insurance). Investors refer to the pursuit of opportunities as "risk and opportunity." The investors are pointing out that one must take on more risk to realize the benefits of the opportunity. In the financial world, decision-making causes risk in the same way that unfortunate events can cause risk. Poor decisions within the organization can prevent its ability to meet objectives. But this is not an event.

With the removal of the word "event," from the definition of risk, it is no longer correct to say that "risk has happened." When there has been an event, it is not proper to say the risk has "occurred." It is also incorrect to characterize a hazard or some other risk source as a risk to characterize a risk as "positive" or "negative." However, it would be valid to describe the consequences of risk as beneficial or detrimental in terms of the organization's objectives.

An event involves the occurrence of a change in a particular set of consequences:

- An event can be one or more occurrences that have several causes.
- An event can consist of something not happening.
- An event can sometimes be referred to as an incident or accident.

DOI: 10.1201/9781003255116-2

- An event can sometimes be referenced as an incident or accident.
- An event without consequences can be referred to as a "near miss."

A consequence is the outcome of an event that affects the objectives:

- An event can lead to a range of consequences.
- A consequence can be certain or uncertain and can have positive or negative effects on the objectives.
- Consequences can be expressed qualitatively or quantitatively.
- Initial consequences can escalate through cascading effects.

An effect is a deviation from the expected and can be positive or negative. Uncertainty has positive effects (opportunities) and adverse effects (threats) concerning an organization's ability to meet its strategic objectives.

The objectives are the overarching outcomes that the organization is seeking to achieve. These effects are the highest expression of intent and purpose and typically reflect the organization's implicit goals, values, and imperatives. Organizations establish responsible objectives; however, to achieve them, they must contend with each operation's internal and external context and all the other organizations in the value chains. Objectives can have different aspects, such as economic, well-being, or environmental. In some organizations, the objectives mirror the three responsibilities of sustainability. We would expect this if the organization was seeking to balance its efforts to achieve sustainability; it must address the stakeholders' interests, achieve its social license to operate, and address risk.

UNDERSTANDING UNCERTAINTY

Uncertainty originates in the internal and external context within which the organization operates. For example, this can be uncertainty that:

- Is a consequence of underlying sociological, psychological, and cultural factors associated with human behavior?
- Is produced by natural processes that are characterized by inherent variability (e.g., weather)
- Changes over time (e.g., due to competition, trends, added information, or changes in underlying factors)
- Is produced by the perception of uncertainty, which may vary between different parts of the organization and among its stakeholders

Uncertainty represents a deficiency of information that leads to an incomplete understanding of what can happen that would threaten the organization's ability to meet its objectives. Think of this as a recession, a severe storm, a devastating legal situation, or any other number of things that could happen that would distract the organization from meeting its objectives. Uncertainty exists whenever the knowledge or understanding of an event, consequence, or likelihood is inadequate or incomplete.

Incompetent knowledge may involve information that alone or in combination with other information:

- Is not available
- Is available but is not accessible
- Is of unknown accuracy
- Is invalid or unreliable
- Involves factors whose relationships or interaction is not known

It may be possible to do something about some of these uncertainty elements, thus lowering the uncertainty; the level of risk is expressed as the likelihood that consequences will be experienced. Consequences relate directly to strategic objectives. Consequences arise when something does or does not happen. Therefore, the possibility of being referred to here is not just that of the event occurring but also the overall likelihood of experiencing the consequences of an event. When uncertainty is present, it creates effects. These effects can lead to a negative or positive deviation from the organization's objectives. Negative effects are often referred to as threats. Positive effects are referred to as opportunities. Risk consists of positive and negative effects.

An organization's objectives must be responsive to its internal and external stakeholders. The practice of organizational sustainability seeks to have the organization establish responsible objectives to help it maintain its social license to operate. Sustainability has always been adept at seeking to find and create opportunities. Risk management is the larger influence and aims to balance the threats and the opportunities. Sustainability is often faulted for being operated in a manner that is not embedded in the organization. The practice of sustainability can identify or create many opportunities to help promote the organization's reputation. However, by working within the risk management program, sustainability can help the organization overcome the effects of uncertainty, thereby enhancing its chance of attaining its strategic objectives.

RISK MANAGEMENT VOCABULARY

All activities of an organization involve risk. Organizations manage risk by coordinating activities to direct and control opportunities and threats regarding the risk posed by each. It can also be seen as the driving force to achieving the organization's strategic objectives.

Risk management will continually be enhanced by people understanding each other's perspectives. This means that everyone needs to know how others view risk. A common issue for large and small organizations is a resource constraint for risk management and control activities. Therefore, it is essential to keep these procedures as straightforward as possible. A risk management system must be a fundamental part of how the organization operates daily and not something only specialists are allowed to work on. Multiple risk definitions and a few different risk management programs are often used within a large parent organization. This presents a problem regarding adopting a common risk language and no overarching program to manage risk effectively.

Interaction between the often-separate risk management fields within an organization (e.g., enterprise risk management, financial risk management, project risk management, safety, and security management, business continuity management, and insurance management) can be ensured or improved, as the attention will not be primarily focused on setting and achieving the organization's objectives, taking risk into account.

Stakeholder perception of risk can also vary to a great degree. This is caused by differences in assumptions, conceptions, and the needs, issues, or concerns related to risk. It makes sense that the stakeholders would seek to make judgments of the acceptability of a risk based on their perception of risk. Therefore, engaging the stakeholders in any risk management program is important. The organization needs to understand their interests and be sure that they are clear about the actual risks involved.

ADOPTING A COMMON RISK LANGUAGE

A common risk language can be created by benchmarking the risk programs to the international definition of risk. This allows different disciplines, units, and geographies with distinct risk profiles to address the unique risks faced by including them in the context description. Risks common to all units in a hierarchical organization are managed strategically. Risks unique to individual, organizational units derive the unit-specific risk responses. It is important to realize that the international definition of risk is high-level. It does not seek to preclude those definitions of the other units.

People advise against using the term risk. Instead, many other terms are used: peril, loss, hazard, threat, harm, danger, difficulty, issue, obstacle, problem, and luck, fortune, accident, possibility, chance, probability, likelihood, uncertainty, consequence, impact, outcome, level, event, occurrence, vulnerability, exposure, benefit, advantage, opportunity, windfall, prospect, and so forth. Even more, reason to have a common risk language! Some practitioners still reject the international standards for risk. However, they often do not recognize that these other words also have a likelihood of misunderstanding when there is communication about risk.

It is crucial to avoid typologies of the areas of risk. Instead, it is more important to focus on understanding the fundamental processes driving uncertainty—hence risk—in organizations. Systems of management, organizational structure, people management, organizational culture power, and conflict all have profound implications for organizational risk, and conflict profoundly impacts organizational risk and uncertainty. The TECOP and PESTLE analyses are helpful when conducted in the determination of the internal and external contexts. These tools help an organization identify uncertainty.

Risk must be considered necessary for the organization's effective strategic planning, management, and decision-making processes. Therefore, evaluating an organization's operating environment is crucial and determining how to integrate risk management with the governance arrangements. The leading edge of risk management practice is addressing the management of uncertainty. This goes beyond perceived threats, opportunities, and their implications. It is about identifying and managing all the many sources of uncertainty that can give rise to and shape the perceptions of threats and opportunities. Uncertainty management implies exploring and understanding the origins of uncertainty before seeking to manage it. There can be no preconceptions about what is desirable or undesirable. Key concerns involve the understanding of where and why uncertainty is essential in each context and where it is not important.

RISK MANAGEMENT FRAMEWORK

A risk management framework is a means of managing uncertainty since the risk is focused on meeting the organization's strategic objectives as seen in Figure 2.1.

The framework does not describe a stand-alone set of activities but rather what is happening within the routine work of the people in the organization. Everything in the framework is fully embedded in how the organization operates daily. It is not a single document or a procedure, even though these elements can be important in the risk management framework. All the risk and uncertainty management activities in any organization can be compared with this framework.

Risk management needs a strong and sustained commitment from the organization's leader to ensure its ongoing effectiveness. Leaders should do the following:

- Create a risk and uncertainty management policy
- Ensure that the organization's culture is aligned with the policy

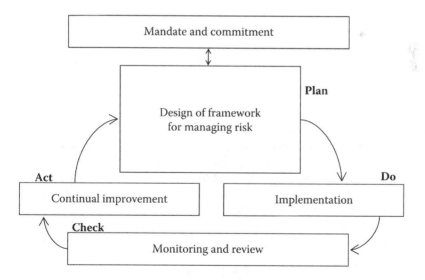

FIGURE 2.1 Design of the risk management framework. (From AS/NZS, Risk management guidelines, companion to AS/NZS ISO 31000:2009, HB 436, SAI Global Press, Sydney, 2013.)

- Monitor and measure the risk management performance in a way that is similar to the measuring of other performance categories
- Embed risk and uncertainty management in the strategic objectives at all levels in the organization
- Align the goals of all members or employees with the risk and uncertainty management objectives
- Ensure legal and regulatory compliance
- Ensure there are proper resources available for risk and uncertainty management
- Communicate the benefits of risk management to all stakeholders
- Ensure that the framework for managing risk and uncertainty remains appropriate

The design of the risk and uncertainty management framework must consider the internal and external context of the organization. This is how it will identify the opportunities and threats (i.e., the effects of uncertainty). Several typical characteristics of the organization need to be considered in the design:

- Organizational structure
- Governing practices
- Policies, internal standards, and operating model
- Contractual requirements
- Strategic and operational systems
- Capability and resources
- Knowledge, skills, and intellectual property
- Information systems and flows

These characteristics should be recorded so they can be referred to from time to time to detect any change that might require the framework to be adjusted.

Members or employees of the organization must have the appropriate competence for managing risk and uncertainty to be held accountable for their role in this risk management framework. The framework must be embedded in all the organization's practices and processes while keeping the contents associated with the framework relevant, effective, and efficient. Many organizations have a risk and uncertainty management plan to ensure that the design elements are embedded in all the organization's practices and processes. Often this plan is part of the strategic plan.

Once the risk and uncertainty framework has been designed, it is time to plan and execute the implementation of the elements so that the risk management process is routinely applied to decision-making throughout the organizations. When implementing risk and uncertainty management, the organization should do the following:

- Define the strategy for implementing risk and uncertainty management
- Apply the risk management policy and process to the organization's processes
- Comply with legal and regulatory requirements

- Ensure the decision-making is aligned with the outcomes of the risk and uncertainty management processes
- Provide awareness development activities
- Engage the stakeholders to ensure that the risk and uncertainty management framework remains appropriate and effective

Any weakness in the design or implementation of the risk management framework can lead to poor performance in meeting the organization's strategic objectives. The framework needs to be monitored and reviewed to determine its effectiveness in supporting organizational performance in the following ways:

- Measure risk management performance
- Periodically measure progress against the risk management plan
- Regularly review whether the design and its components are still appropriate
- Report on the results of monitoring and reviews; decisions should be made on how the risk and uncertainty management framework, policy, and plan can be improved. These decisions should improve the organization's management of risk and its risk management culture.

Based on the results of monitoring and reviews, decisions should be made on how the risk and uncertainty management framework, policy, and plan can be improved. These decisions should improve the organization's management of risk and its risk management culture.

EMBEDDING RISK MANAGEMENT

The focus of every organization should be on meeting its strategic objectives. This starts with setting responsible objectives that cascade from the top to the bottom of the organization. At this point, goals are set along with action plans to make sure the objectives are met. Risk management should be embedded in the activities to manage the opportunities and threats associated with the effects of uncertainty in the internal and external operating environment. Objectives must be established and maintained, mindful of these opportunities and threats. The people responsible for setting the objectives should also be responsible for managing risk. However, the leaders of an organization know that everyone is accountable for risk management within the structure of objectives and goal setting. Risk management should not be delegated to specialized risk practitioners in separate departments. Support staff plays a crucial role in assisting the "owners" with the effective management of risk.

All the functional units in an organization (e.g., environmental, health, safety, assets, quality, human resources, legal, sustainability, purchasing, and communications) must not have objectives that cannot be traced to the organization's strategic objectives. Therefore, uncertainty should always be identified, assessed, responded to, reported, monitored, and reviewed in relation to the organization's objectives, considering changes occurring in the internal and external context. When the members of an organization begin to understand the effects of uncertainty on the opportunities and threats, they can seek the guidance of the organization's governance to manage the risk associated with this uncertainty. Also, risk needs to be considered before a decision is made or actions have been initiated. This is true no matter where a decision is made within the organization.

Risk management must be tailored to the organization. Parent organizations must realize differences in the internal and external contexts at each facility or operation. The risk is like strategy. Both must be adapted to ensure success at the application stage. Where risk management is embedded within the governance of the organization, the risk management function plays several essential roles:

- Facilitates proper risk management and internal control processes within all levels of the organization
- Serves as the custodian of the overall risk management and control frameworks
- Provides internal assurance on effective risk management and internal control within the origination

The effects of uncertainty can never be eliminated. Therefore, organizations need to build resilience and agility in all their activities that enable them to respond to changes in circumstances or deal with the consequences of unforeseen events. All eyes need to be on the opportunities and threats and the uncertainty analysis used to prioritize them whenever there is a change in the internal or external context.

Risk management is built into organizations' systems of management. Risk elements can be added or eliminated as needed. Once risk management is fully embedded as an integral part of the management system, the organization's management system helps the members or employees make intuitive decisions and take responsibility and sustainable actions.

WORKING WITH RISK AND RISK MANAGEMENT

The *Oxford Dictionary* defines "risk" as a situation involving exposure to danger. Asking people for their definition of risk provides a wide variety of definitions. Some of their responses focus on a concern for uncertainty or danger, while others refer to the financial consequences of unwanted events. Every organization faces some degree of risk every day. However, we usually focus on catastrophic events and whether the organization is adequately covered by insurance protecting us from the consequences of these events.

Other sources of risk include the following:

- The possibility of an unfortunate occurrence
- Doubt concerning the about come of a situation
- Unpredictability
- Possibility of a loss
- Needed to improve the ability to be an effective leader

These concepts of risk help us understand risk broadly as the uncertainty of future events and their outcome for our organization.

Leading companies create or adopt frameworks for understanding risk and supporting risk management. Typically, the approach to understanding risk supports the business and its external context while ensuring that risk management is embedded across the entire organization. Leading companies create or adopt frameworks for understanding risk and supporting risk management. Typically, the approach to understanding risk supports the business and its internal and external context while ensuring that risk management is embedded across the entire organization. This action requires an explicit management dialogue with every element of the organization and its key stakeholders. Corporations do not like risk or uncertainty. In these organizations, new initiatives are carefully reviewed to either eliminate risk or mitigate that risk to levels acceptable to the organization. This situation makes these companies more vulnerable to disruption as entrepreneurial companies tend to put risk aside or accept a higher risk tolerance to impact how organizations conduct their business.

As a result, organization leaders need to understand risk and uncertainty. There are manageable ways to understand risk without confusion by all the risk-naming conventions. The organization should thoroughly search for risks as the first step in a risk management program. This list must be updated whenever company changes occur or circumstances relevant to the organization change. It is not necessary to build a complicated risk classification system. The major risk management program standards do not encourage the classification of risk. A few critical concepts necessary to understand risk are presented below.

PURE RISK AND SPECULATIVE RISK

A pure risk (traditional risk) features a chance of a loss and no chance of a gain. People often use the word "risk" to describe a financial "loss." Losses result from fires, floods, snow, hurricanes, earthquakes, lightning, and volcano eruptions. The business's losses include more complex matters such as sickness, fraud, environmental contamination, terrorism, electronic security breaches, and strikes. A risk is the possibility of a loss. A peril is the cause of a loss. Perils expose people and property to the risk of damage, injury, or loss against which the organization often purchases insurance to cover the cost of that loss. Please note that the terms "peril" and "loss" are often mistakenly used interchangeably.

Insurance companies cover financial losses from pure risks that meet conditions: chance, definitiveness, and measurability; statistical predictability; lack of catastrophic exposure; random selection; and loss exposure.

Pure risk involves a hazard. A hazard is something that increases the probability that a peril will occur. Hazards are a condition or a situation that makes it more likely that a peril will occur. The situations include physical hazards, operational hazards, and business hazards. Common hazards include chemicals, repetitive motions, and physical conditions (e.g., vibrations, noise, slips, trips, falls, ergonomic situations, and biological effects).

Speculative risks are activities that produce a profit or a loss. These activities include new business ventures, reputation protection, modifications to operations, and alternative means of transportation. All speculative risks are undertaken as a result of a conscious choice. As a result, speculative risk lacks many of the core elements of insurability.

FINANCIAL AND NON-FINANCIAL RISK

Larger corporations focus on the financial risk of their operations; financial risk refers to an organization's ability to manage its debt and financial leverage. It also refers to non-debt financial losses, like litigation, property losses, crime, fraud, and cyber risk. Organizations create performance measures to address these financial risks, including cash flow, credit earnings, equity, foreign exchange, interest rates, liquidity, and financial reporting.

However, to have a vibrant risk management program in an organization, it is important to consider the non-financial risks associated with the operations. Non-financial risks are events or actions other than financial transactions that can negatively impact the operations or assets of a company. Typical non-financial risk includes misconduct, technology, and ignoring key external stakeholders, customers, and employees.

However, there are some drawbacks associated with non-financial performance measures. First, monitoring a large amount of financial and non-financial information is costly. In some cases, the cost is greater than the benefits. Having many performance measures requires maintaining and studying information from multiple sources. There is often a competition between maintaining a good set of measures and finding the time needed to spend more time engaging with stakeholders and serving the customers.

There are established and certifiable means of measuring financial performance. However, this is not the case in non-financial measures. Evaluating performance or making trade-offs between measurements is complex when some are measured in time, others are measured in percentages or amounts, and a few are determined arbitrarily. Furthermore, all stakeholders understand or hold a similar appreciation of non-financial measures. Lastly, accounting systems are designed around financial measures and do not handle non-financial concepts well.

Although non-financial measures are receiving more attention in risk management programs, organizations should not simply copy the measures from other organizations. Instead, the choice of the non-financial measures should be unique to each company and linked to organizational strategy and meeting the organization's explicit objectives and other value drivers.

OPPORTUNITIES AND THREATS

With the advent of the practice of risk management (i.e., different than hazards control) in the 1990s, there has been a shift to using opportunities and threats to manage an organization's risks. As noted earlier, the traditional view of risk is negative. This view characterizes all risks as "threats" with adverse consequences on the ability of the organization to meet its objectives. However, there is a possibility that uncertainty in the internal and external operating environments can create an "opportunity" that has a beneficial effect on achieving organizational objectives. This is consistent with the more receipt view of risk as being the "effect of uncertainty" on the ability of the organization to meet its business objectives. The nature of uncertainty and its effect on objectives can change over time. As a result, the risk will vary. What is found in uncertainty today may not be accurate in the future. Since most business strategic objectives are established for five to ten years, monitoring and measuring the operating environment is imperative.

The International Organization for Standardization (ISO) defines effect (as in the effects of uncertainty) as "a deviation from the expected positive or negative."[7] Opportunities brought to light are often not the opportunities that might have been already known to the organization. They are challenging opportunities, so the thought of them being a "risk" is very apropos. Most organizations that are using the opportunities and threats in their risk management program select a couple of the highest-ranked opportunities and seek to exploit them instead of simply treating the top threats that have been identified. It is important to remember that you should not use the word "risk" interchangeably with the word "threat."

CONTEXT RISK

The context risk is defined as follows:

> The effect of uncertainty on objectives. An effect is a deviation from the expected. It can be positive, negative, or both. An effect can arise from a response or failure to respond to an opportunity, or a threat related to objectives. Risk is usually expressed in terms of risk sources, potential events, consequences, and likelihoods.

Establishing the context of an organization is concerned with understanding the external and internal operating environments to identify the risks (i.e., opportunities and threats) that would be of concern to the company. In addition, the information obtained from determining the context risk will help identify the structure for the risk management activities. Therefore, careful delineation of the context risk is needed to:

- Clarify the organization's objectives
- Identify the operating environments within which the objectives are pursued
- Specify the scope and objectives for the risk management boundary conditions and the outcomes
- Identify the criteria that will be used to measure the risks
- Define a set of critical elements that will be used to structure the risk identification and assessment process

The context establishes, implements, maintains, and continually improves the organization's high-level structure thus creating an understanding that could exert an influence on the ability of an organization to meet its objectives (i.e., outcomes). The organization determines those opportunities and threats that must be addressed and managed.

RISK AND RISK MANAGEMENT VOCABULARY

When dealing with risk in the context of an organization, it is essential to share a common language regarding risk and risk management. Company communications concerning risk management efforts must use an agreed-upon vocabulary. The top leader should ensure that these terms are consistently used within the organization and when seeking stakeholder engagement. The International Organizations for Standardization (ISO) has created an open-source document.

Hazard risks undermine objectives and often have a high significance in some industries. These hazard risks are closely related to insurable risks. Remember that a hazard (or pure risk) can only have a negative outcome. The occupational health and safety management system ISO 45001 is very careful in maintaining information on both hazard risk and risk associated with the effects of uncertainty. Consider the wording in this standard's Section 6.1.2.2.

> The organization shall establish, implement, operate, and maintain the OH&S management system.[8]

All management systems will need to separate the hazard risks similarly.

ESSENTIAL QUESTIONS FOR ORGANIZATIONAL RISK MANAGEMENT

1. Define risk. How does traditional risk differ from organizational risk?
2. What are the critical areas in which the organizational leader must have a strong and sustained commitment to managing uncertainty to meet the organization's strategic objectives?
3. What word should replace the word "risk" in organizations? How does this change the perception of risk?
4. What are opportunities, and how do you find them when examining risk in an organization?

REFERENCES

1. ISO (International Organization for Standardization), "Guidelines for addressing sustainability in standards", Guide 82, Geneva, 2014.
2. R. Pojasek, "Understanding sustainability," Understanding sustainability: An organizational perspective, *Environmental Quality Management*, p. 93–100, 2012.
3. R. Pojasek, "Sustainability: The three responsibilities," *Environmental Quality Management*, Cambridge, p. 87–94, 2010.
4. Standards Australia International, "Organizational code of conduct. AS 8002-2003," Sydney: Standards Australia International Ltd, 2003, [Online]. Available: https://www.saiglobal.com/PDFTemp/Previews/OSH/as/as8000/8000/8002-2003(+A1).pdf

5. ISO (International Organization for Standardization), *Risk Management—Guidance for the Implementation of ISO 31000*, Geneva, 2014.
6. ISO (International Organization for Standardization), *Social Responsibility Guidance. ISO 26000*, Geneva, 2010, [Online]. Available: https://www.iso.org/publication/ PUB100258.html .
7. ISO (International Organization for Standardization), *ISO 37301:2021 Compliance management systems — Requirements with guidance for use*, Geneva, 2014.
8. Geneva, *Health and Safety Management Standard ISO 45001*, 2018.

3 Decision-Making

ABSTRACT

Organizations achieve sustainable success in part by improving the ability of all leaders, members, or employees to make decisions. In most cases, poor decisions are more of a threat to organizations than events. The success of any sustainability effort depends on the effective use of decision-making, knowledge management, and sense-making. However, sustainability professionals cannot help the organization improve its decision-making by operating remotely. When sustainability is embedded within the organization, its priority should be to understand the decision-making process and instill the three sustainability responsibilities and principles into this vital process.

DECISION-MAKING IN ORGANIZATIONS

Leaders in an organization need to understand the effects of uncertainty (opportunities and threats) to create a decision-making process that helps the organization meet its strategic objectives. Most people in an organization are involved in decision-making, so the process must be clearly defined and carefully monitored within a sustainability program. When an event threatens the organization, the uncertainty is heightened, and the organization is likely to be on the downside of risk. On the other hand, the event provides the organization with an opportunity that can be realized to help the organization be on the upside of risk. In these cases, an event is an occurrence or change of a particular set of circumstances.[1] The event could be one or more occurrences with several causes or something that did not happen.[1] However, it does not take an event to influence the risk. Therefore, the significant opportunities and threats are addressed with decisions involving no event.

Making decisions is part of the risk and sustainability management processes. Making good decisions can ensure the organization's ability to meet its objectives. Risk is created or altered when decisions are made because uncertainty is almost always associated with decision-making. Risk is always associated with strategic objectives. Everyone must understand that risk-taking is unavoidable in all the organization's activities. The risk associated with a decision should be understood when the decision is made, not after the decision has already been made. Risk-taking must be intentional.[2]

Risk management provides the foundation for informed decision-making. It needs to be integrated into activities supporting the achievement of the strategic objectives and the decision-making process.[2]

DOI: 10.1201/9781003255116-3

- Decisions made on the organization's strategic issues should consider any uncertainties (opportunities and threats) associated with the external operating environment and changes in the organization's internal context.
- The organization's innovation process should consider not only the identified uncertainties but also uncertainties related to the innovation's human, social, safety, and environmental aspects, and be managed according to legal requirements, when applicable.
- Plans for significant financial investment should specify the decision-making milestones at which uncertainty analysis will occur.
- The organization's policy on risk management should reflect these and other points that influence the ability to meet the strategic objectives. There needs to be a framework that guides decisions to apply the process effectively and consistently in all decision-making. This requires a clear allocation of accountability, supported by skill development and performance review.[2]

DECISION-MAKING PROCESS

In an ideal situation, decision-making requires a complete search of all available alternatives, reliable information about their consequences, and consistent preferences to evaluate the potential outcomes. Unfortunately, this never happens in the real world. Everyone is limited by their cognitive and mental capabilities, the extent of knowledge and information available, and conflicts of interest that are not aligned with the organization's objectives. Nevertheless, organizations usually have a decision-making framework that helps control decision-making.[3] Decisions are made at all levels of the organization. They can be:[4]

- Strategic: Related to the design of a plan of action to achieve an objective
- Tactical: Related to the way different parts of an organization are arranged to deliver the objective
- Operational: Related to the way individuals work daily to accomplish specific results—their goals

Organizations that make better, faster, and more effective decisions will have a competitive advantage.[5]

Decisions are where thinking and doing overlap. For this to happen within an organization, a decision must be logically consistent with the organization's risk management program and the willingness of the operations to agree that they can do what is decided. Therefore, decision-making involves a defined process and should be handled this way.

Decision-making is connected with sense-making and knowledge management (Figure 3.1).

It is focused on helping the organization meet its objectives in an uncertain world. Decision-making uses knowledge to improve its ability to have successful outcomes. It also records what is learned during the decision-making process in the knowledge management system. Finally, sense-making takes the information from scanning the

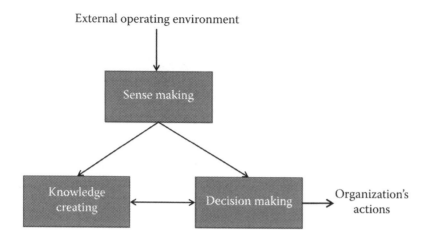

FIGURE 3.1 Relationships between decision-making, sense-making, and knowledge management. (Adapted from Choo, W.C., The Knowing Organization: How Organizations Use Information to Construct Meaning, Create Knowledge and Make Decisions, 2nd ed., Oxford University Press, New York, 2006.)

external operating environment and either stores it in the knowledge management system or tries to make sense of the information and make it available within the decision-making process. These two additional elements will be discussed in the following sections.

DECISION-MAKING TECHNIQUES

There are many different kinds of decision-making techniques. Organizations select the methods depending on styles that range from autocratic to unanimity-based decision-making.[4] The criteria that shape decision-making can be stated as follows:[4]

- A decision environment that may influence the decision style
- Complexity of the decision
- Value of the decision's desired outcome
- Alternative scenarios that have the potential to support the decision-making process
- Cognitive biases to the decision's selection and interpretation
- Quality requirements of the decision
- Personalities of those involved in decision-making
- Time available to conduct the decision-making process
- Necessary level of commitment to or acceptance of the decision
- Impact on valued relationships that the choice of decision style may have

Several of these criteria may be in play at any moment and amplify one another.

Decisions should be made by following a sequence of specific steps:[4]

- Organizations should know which decisions have a disproportionate impact on organizational performance.
- Decision-makers should determine when those decisions should happen.
- They should organize the structure of decision elements around sources of value for the organization.
- Decision-makers should figure out what level of authority is needed, regardless of status, and elevate it to that level.
- They should align other parts of the organizational system, such as processes, data, and information, to support decision-making and execution.
- Decision-makers should help managers develop the skills and behaviors necessary to make decisions promptly and diligently translate them into action.

An organization's value is often seen as the sum of the decisions it makes and executes. However, even good decisions can have unfavorable results. Most organizational decisions are not made in a logical, rational manner. Most decisions do not begin with careful analysis, followed by systematic analysis of alternatives and implementation of a solution. On the contrary, decision processes are usually characterized by conflict, coalition building, trial and error, speed, and mistakes. Leaders operate under many constraints that limit the rationality of their decisions. As a result, they use intuition as well as rational analysis in their decision-making.[6]

WORKING WITH LIFE CYCLE THINKING (LCT)

Life Cycle Thinking (LCT) is a powerful tool to gain a 10,000-foot perspective about your organization's resources, energy, and water consumption by making your products or services. Life cycle thinking looks at all inputs and outputs of that product in various stages of its life. To understand the life cycle perspective of their inputs and outputs, an organization does not need to perform an actual life cycle analysis of their products but can understand their impact through a life cycle thinking process. First, inputs are resources consumed, which can be raw materials such as wood, water, or electricity. Outputs are emissions from the creation of the electricity or waste generated when you cut the wood. Next, life cycle thinking covers the production stages that happen because of producing the organization's marketable product or service. Each stage is interconnected to many other stages in the process, both directly and indirectly. Typical stages of a product (or service) life cycle include:

- Raw material acquisition
- Transportation
- Production or manufacturing
- Use
- End of life

The inputs and outputs of the product are measured from the first phase of raw material extraction, through transportation to the manufacturing or processing facility, the manufacturing process, the use phase, and finally, as the product reaches its grave or final disposal. In this approach, only the inputs and outputs of resources and waste are considered for a specific life cycle analysis product, but in life cycle thinking, more "non-traditional" impacts can be regarded as social equity and stakeholder interests.

LCT is a comprehensive view of a life cycle accounting for economic, environmental, and social impacts across all stages of a product or process. LCT is a significant first step in discovering the scope of where you should look to find sustainability targets. It can narrow your range anywhere in operation to specific locations targeted by resource consumption. Every time you spend money to consume resources, you can save by becoming more efficient and reducing waste. Understanding what is in your products and where the materials come from is a massive step for some organizations.

Start the step-by-step process of LCT for resources first in an organization. The resource-first approach allows an organization to find its "low-hanging fruit" and, therefore, by becoming more efficient, save money. Efficiency is also relatively easy, gets employees engaged, and creates quantifiable "wins" for the initiative. To start the process, write out the life cycle stages: raw materials, transportation, production, distribution/warehousing/retail (if applicable), customer use, and end of life. The organization will have the most significant operational control from the middle of the process (manufacturing) back to raw material extraction, with its operational control lessening with each step backward. For instance, a manufacturer can influence the types of raw materials they purchase through sustainable procurement, but they are limited by the information their suppliers share. As a result, there may be non-sustainable attributes they are unaware of that the supplier has yet to disclose. This contrasts with the manufacturer's operations, where they control every detail.

The same waning influence is true for the steps from the middle (manufacturing) forward to some degree, but while the organization may have some control over its distribution/warehousing/retail, its operational control is at its weakest in determining what the customer does at the end of the life of the product or service. Next, connect all the stages. This is mostly transportation and will illustrate where fossil fuels are consumed. Finally, the organization should consider those stages in the life cycle over which it has the most significant control or influence. These may offer the greatest opportunity to reduce resource use and minimize pollution or waste, which is easily qualifiable and saves the organization's bottom line. This is an example of the perspective of LCT, and when considering all the stages listed above, it is Cradle-to-Cradle life cycle thinking.

From an economic perspective, an organization can set efficiency-based targets that align with its yearly objectives. But there are other objectives an organization needs to consider in their evaluation; therefore, the perspective of the LCT map can be refined and changed based on the objective considered. An example of a specific LCT perspective is examining risk in the supply chain due to shipping delays. Mapping out the primary raw materials needed to produce a product, the distance from the extraction/processing site to the manufacturer, and by which mode of transportation it arrives at the facility and then comparing that to the lead times needed for manufacturing that item could highlight huge hotspots of risk in the life cycle of that product. This type of LCT thinking is called Cradle-to-Gate because it is only concerned with the beginning of the extraction process through the point the product is sitting on the dock waiting to be shipped to the distributor, retail location, or customer has a shipping time-based perspective.

A significant benefit of LCT is developing a plan for resource efficiency, but beyond that, LCT can contribute to organizational factors internally and externally. Using Life Cycle Thinking, organizations can internally improve productivity and product quality, reduce costs, improve employee engagement, and foster innovation. Organizations can use LCT to enhance their image by selecting sustainability targets stakeholders care about, driving a market advantage, gaining goodwill with regulators, and positively strengthening their license to operate in a community. All these factors combined equal, reducing risk and increasing their competitiveness. By focusing on the internal driving forces, an organization develops the knowledge, commitment, and readiness to readily embrace new trends and establish its organization as a market leader because it strategically develops and markets better products.[7] And it all starts with understanding the entire life of their product or services.

Life Cycle Thinking is an integrative tool invaluable for creating product-centric sustainability targets or entire organizational sustainability programs. The difference is in the LCT perspective and depth of study. Overall, the organization can be flexible in its scope and only include the environmental aspects that it can control directly, or they can be related to products and services used by the organization which is provided by others, as well as products and services that it offers to others, including those associated with their supply chain and any OEM suppliers. The tool's power is the flexibility of the perspective to find hot spots specific to the organization and use that information to make decisions on them.

Since life cycle thinking maps the current state of "what is," it can contribute to the decision-making process and help determine the upstream and downstream consequences before project implementation. When you make improvements in one area based on information gathered during the LCT process, you can also assess alternative scenarios, develop and select alternative implementation strategies, and develop indicators for monitoring and evaluating outcomes. LCT in decision-making can also highlight potential pitfalls in the plan by considering the action from multiple viewpoints.

Stakeholders can also be layered onto life cycle thinking as part of the decision-making process. Mapping stakeholders and related activities and the environmental impacts in the LCT process will enable the stakeholders to expand their thinking and, therefore, potentially expand their contributions toward the issues that may promote or hinder the sustainable target the organization is working toward. The impacts could be qualitative or quantitative, depending on data availability and the nature of the effects. This, in turn, could encourage stakeholder commitment and cooperation toward the sustainability project.[8]

PREPARING TO MAKE A DECISION[9]

When preparing to make a decision, the decision-maker(s) must consider the following:

- Understand the purpose of what the organization is trying to achieve. What will the decision involve?
- Define the outcomes that would be desirable and their relation to helping the organization achieve its strategic objectives. Know how the outcomes will be measured.
- Consider who needs to be involved in the decision and the actions that will follow.
- Determine the criteria that will be used to make the decision. For example, what is the basis for knowing whether the decision will be correct considering the strategic objectives?
- Determine how to decide by defining the process and steps. For example, what will be considered, and what will be deemed irrelevant?

MAKING THE DECISION[9]

Making the decision is like playing chess, the decision-maker must first seek to understand the playing field, as in the step above, but in making the decision they must consider the intended and unintended outcomes. When making decisions consider the following:

- Seek clarification and agreement on the key assumptions about the decision and the expected outcomes.
- Consider the range of things that might occur or are already present that would prevent achieving the intended outcomes.
- Do the same for the case of enhancing the achievement of the objectives.
- Consider the effect of those things that might occur or are already present on achieving the organization's strategic objectives.
- Reflect on the decision-maker's recent performance under similar circumstances. Does this indicate the effectiveness of existing operating controls that are intended to enable the achievement of the organization's strategic objectives?
- What lessons can be learned by studying any recent activities that are similar to the focus of the current decision?

ACTING AFTER THE DECISION[9]

Once the decision has been made, an action plan should be put into place to measure the effectiveness of the outcomes. Consider

- Limiting uncertainty associated with the desired outcomes by identifying new opportunities and threats and managing them

- Costs and benefits of alternatives to the path selected and decide which creates the most value and provides the highest chance that the outcomes will be achieved
- The critical path to achieving the desired outcomes and allocating resources and accountabilities for all actions

In the end, ensure that the decision-making and planning process is captured in documented information and maintained within the knowledge management system. It is an essential step because the decision-making process should not be ad hoc and should be established to measure performance and ensure the completion of all the planned actions.

MONITORING AND REVIEWING ACTIONS[9]

Decision-making is a complex process. Track the progress and actions with appropriate oversight. Question any deviances from what was planned. Reallocate resources if required. As part of the continuous review process, conduct periodic reviews to monitor:

- The opportunities and threats (uncertainty) and scan the internal and external operating environment, if necessary, to see if there is additional uncertainty. Use sense-making to determine the implications of any changes on what the organization is trying to achieve and its strategic objectives.
- The existing controls intended to enable the organization to achieve its strategic objectives
- Decision-making and implementation as part of the organization member's or employee's performance evaluation while encouraging sound decision-making

Lastly, verify continuous monitoring and periodic reviews with some level of internal audit.

LEARNING IN DECISION-MAKING[9]

Learning is an important stage in the decision-making process. First there must be an agreement on the extent to which the desired outcomes were achieved. Additionally, all stakeholders involved in the decision-making process must agree on whether:

- The outcomes helped or hindered the ability of the organization to meet its strategic objectives.
- Any consequential effects were successes or failures.
- The causes of the successes or failures and the roles of operational controls.

Once the stakeholders agree, then consider the implications for future decisions and define what the organization should learn. Try to capture all the information and learning in the knowledge management system.

SENSE-MAKING

Sense-making monitors changes in the external operating environment (context). It seeks to make sense of what is happening, why, and what it means. Sense-making is always performed retrospectively because it is impossible to make sense of events and actions until they have occurred and there has been sufficient time to construct their meaning concerning decision-making in the organization.

The sense-making process is usually put into play when some change occurs in the external operating environment. Through this process, the organization attempts to understand the differences and determine the significance of those changes. There are three basic steps involved in sense-making:[3]

- Enactment: The information is labeled, categorized, and connected with information about stakeholders, events, and outcomes. Sometimes, the organization introduces new messages or actions into the external operating environment (e.g., distributing a document, holding a meeting, or creating a website) and then refocuses its sense-making on these activities. The enactment process segregates possible factors that the organization should clarify and take seriously.
- Selection: People look at the information gathered from the enactment and try to answer the question, what is going on here? This can be accomplished by overlaying new data with interpretations that have worked in explaining similar or related situations that have happened before. First, interpretations are selected that provide the best fit with past understandings. Then, a set of cause-and-effect explanations are created that could explain what was observed. These explanations have to be plausible, but they do not need to be the most accurate or the most complete.
- Retention: The products of successful sense-making are retained in the knowledge management system for further use in the future.

Sense-making is used to produce reasonable interpretations of data about the change in the external operating environment of an organization. It can be used in conjunction with scanning the external operating environment. Using the PESTLE tool, together with the enumeration of the opportunities and threats, produces valuable information in the sense-making process. In addition, engagement of the stakeholders in the external operating environment offers opportunities for collaboration in both the scanning and the sense-making activity. In times of rapid change, organizations must take advantage of these tools to help deal with the effects of uncertainty (opportunities and threats).

KNOWLEDGE CREATION AND MANAGEMENT

All organizations need to develop the capacity to create new knowledge continuously. Knowledge creation involves the management of tacit and explicit knowledge. It works best when a process generates new knowledge by converting tacit knowledge into explicit knowledge. Tacit knowledge is the personal knowledge that members of the organization carry around in their minds. It is hard to formalize or communicate this knowledge to others. It consists of subjective "know-how," insights, and intuition derived from performing tasks over a long period. Explicit knowledge is formal knowledge that is easier to transmit between individuals and the organization. It may be codified in work instructions, procedures, operational controls, rules, and other forms.[3]

The production of knowledge involves the conversion of tacit knowledge into explicit knowledge and back again. This typically takes place in four steps:[3]

1 Socialization is a process of acquiring tacit knowledge through sharing experiences. Tacit knowledge is transferred from one experienced person to another by working side by side. This is built on the long-standing tradition of apprenticeships.
2 Externalization is a process of converting tacit knowledge into explicit concepts through the use of abstractions, metaphors, analogies, or models. This is what most people think of when talking about knowledge creation. It is often used in the concept creation phase of new product development.
3 Combination is a process of creating explicit knowledge by finding and bringing together detailed expertise from several sources. In this case, individuals exchange and combine their explicit knowledge through telephone conversations, meetings, emails, and other media forms. In addition, existing information in computerized databases may be categorized, collated, and sorted to produce new explicit knowledge.
4 Internalization is a process of embodying explicit knowledge into tacit knowledge, internalizing the experiences gained through the other modes of knowledge creation into individuals' tacit knowledge assembled in the form of shared work practices.

These four modes of knowledge conversion follow each other in a continuous spiral of knowledge creation.

TRANSITION TO A LEARNING ORGANIZATION

Outcomes of sense-making are enacted environments or shared interpretations that construct the organization's context. The knowledge-creating model (Figure 3.1) depicts the organization continuously tapping its knowledge to solve challenging problems. Different forms of organizational learning are converted and combined in continuous innovation cycles. The outcomes of the process include the development of new products and services, as well as new capabilities. The decision-making model sees the organization as a rational, objective-directed system. Decision-makers search for alternatives, evaluate consequences, and commit to a course of action using the gathered information from sense-making and knowledge creation. The outcome of decision-making is the selection of courses of action intended to enable the organization to achieve its strategic objectives.[3]

Even while the rational decision-making framework is probably widely used by a wide range of organizations, it is common for people to gather information for decisions and not use it. They ask for reports but do not read them. Individuals fight for the right to participate in decisions but do not exercise that right. Policies are vigorously debated, but their implementation is met with indifference. Leaders spend little time making decisions because they are engaged in meetings and conversations about pending decisions.[3]

The knowing cycle model provides a structure and language that can be used to think about the use of information in organizations. The model does not specify a particular sequence or order (Figure 3.1). Instead, the three processes are simply interconnected, with many possible pathways along which the model can function. First, sense-making constructs the context, the frame of reference for knowledge creation and decision-making. Knowledge creation expands organizational capabilities and introduces innovation. Finally, decision-making converts belief and capabilities into commitments to act.[3]

LEARNING IN THE ORGANIZATION

The organization should encourage improvement and innovation through learning. For the organization to attain sustained success, it is necessary to adopt "learning as an organization" and exploit learning that integrates the capabilities of individuals with those of the organization as a whole.[10]

Learning as an organization involves consideration of the following:[10]

- Collecting information from various internal and external events and sources, including success stories and lessons learned
- Gaining insights through in-depth analyses of the information that has been collected

Learning that integrates the capabilities of individuals with those of the organization is achieved by combining people's knowledge, thinking patterns, and behavioral patterns with the organization's values. This involves the consideration of:[1]

- The organization's values are based on its mission, vision, and strategies
- Supporting the development of learning models and demonstrating leadership through the behavior of leaders
- Stimulation of networking connectivity, interactivity, and sharing of knowledge both inside and outside of the organization
- Maintaining systems for learning and sharing knowledge
- Recognizing, supporting, and rewarding the improvement of people's competence through processes for learning and sharing knowledge
- Appreciation of creativity supporting a diversity of the opinions of the different people in the organization

Rapid access to and use of such knowledge can enhance the organization's ability to manage and maintain its success. Learning better and faster for organizations wanting to remain relevant and thrive is critically important. Organizational learning is only possible and sustainable with an understanding of what drives it.[5]

A learning organization values the role that learning can play in developing organizational effectiveness. It demonstrates this by having an inspiring vision for learning and a learning strategy that will support the organization in achieving its strategic objectives.[5]

A learning organization needs intellectually curious people who actively reflect on their experience, develop experienced-based theories of change and continuously test these in practice with colleagues, and use their understanding and initiative to contribute to knowledge development. Reflective practitioners understand their strengths and limitations and have a range of methods and approaches for knowledge management and learning, individually and collectively.[5] Knowledge is a critical asset in every learning organization. Because learning is both a product of knowledge and its source, a learning organization recognizes that the two are inextricably linked and manages them accordingly.[5]

Learning organizations know how to harness the power of information and communication technologies without these technologies constraining knowledge management and learning. In a learning organization, information and communications technologies are used, among other purposes, to strengthen organizational identity; build and sustain learning communities; keep members, clients, customers, and others informed and aware of organizational developments; create unexpected, helpful connections between people and provide access to their knowledge and ideas; encourage innovation and creativity; share and learn from good practices and unintended outcomes; strengthen relationships; develop and access organizational memory; share methods and approaches; celebrate successes; identify internal sources of expertise, and connect with the outside world.[5]

It is essential for organizations that wish to embrace sustainability to build a learning organization. However, isolated knowledge management initiatives will only last for a while, irrespective of the approach to creating a learning organization. Only embedded, organization-wide activities to identify, develop, store, share, and use knowledge can give knowledge management the role to help with decision-making and provide part of the foundation necessary for the sustainability journey.[4] Despite the attention paid to strategic planning, the notion of strategic practice is relatively new. Therefore, creating a strategy is relatively easy. However, executing that strategy is quite tricky since strategy needs to be synchronized with sense-making, knowledge management, and decision-making. Therefore, organizations should systematically review, evaluate, prioritize the sequence of, manage, redirect, and, if necessary, cancel strategic initiatives in a learning organization.[11]

ESSENTIAL QUESTIONS FOR DECISION-MAKING

1. What are the key steps in the decision-making process? Which one is most important, in your opinion, and why?
2. What is Life Cycle Thinking (LCT) and how can it be used in organizations as part of the decision-making process?

3. How do risk management and LCT both contribute to the decision-making process in an organization?
4. Why should an organization transition to a learning organization?

REFERENCES

1. ISO (International Organization for Standardization), *Risk Management—Principles and Guidelines*, Geneva, 2009.
2. ISO (International Organization for Standardization), *Risk Management—Guidance for the Implementation of ISO 31000*, Geneva, 2013.
3. W. Choo, "The knowing organization: How organizations use information to construct meaning, create knowledge and make decisions," *The Knowing Organization*, 2nd edition, New York, 2006.
4. O. Serrat, "On decision making," *Knowledge Solutions*, Manila: Asian Development Bank., 2012.
5. O. Serrat, "On organizational configurations," *Knowledge Solutions*, Manila: Asian Development Bank, 2012.
6. R. Daft, *Organization Theory & Design*, Mason, OH: South-Western Learning, Centage Learning, 2013.
7. Danish Environmental Protection Agency, *An Introduction to Life Cycle Thinking and Management*, Denmark: Danish Ministry of the Environment, 2003.
8. A. W. R. R. Lanka Thabrewa, "Environmental decision making in multi-stakeholder contexts: Applicability of life cycle thinking in development planning and implementation," *Journal of Cleaner Production*, pp. 61–67, 2009.
9. V. Tophoff, "Managing risk as an integral part of managing an organization," *From Bolt-on to Built-in*, 2015.
10. ISO (International Organization for Standardization), *Risk Management: Principles and Guidelines, ISO 31000*, Geneva, 2009.
11. Serrat, Olivier, and Olivier Serrat, "Understanding complexity," *Knowledge Solutions: Tools, Methods, and Approaches to Drive Organizational Performance*, 2017: 345–353.

4 Organizational Objectives and Leadership

ABSTRACT

To achieve sustainable success, an organization must meet its overarching objectives over the long term while operating in an uncertain world. When developing a sustainability program, it is essential to understand its objectives and how they were established. There is disagreement regarding the difference between "objectives" and "goals" "as with other terms." Many people continue to use these terms interchangeably. There is also disagreement regarding the purpose and content of a mission statement—the principal source of the organization's objectives. This chapter focuses on setting objectives and goals in an organization.

ORGANIZATION'S MISSION STATEMENT

The mission statement is widely regarded as an explicit statement of the reason for the organization's existence and what it is meant to accomplish. Mission statements are typically focused on a five- to ten-year time frame and should:

- Separate what is vital to the organization's sustainable success from what is not as important
- Clearly state the customers, clients, or other persons and organizations that are served and how they are served
- Communicate the organization's "looking forward" position to its stakeholders

The mission is focused on its products or services for a private business. In the public sector, the mission statement focuses on what the organization is trying to accomplish. Most not-for-profit organizations are established to start something new or stop something that they find objectionable, such as protecting a wetland area or eliminating the emission of greenhouse gases. In all cases, the mission statement provides a foundation for the strategic planning process and the management of the processes and activities within the organization. A mission statement is significant to an organization because it helps management achieve sustainable success over time.

A mission statement is different from a vision statement. Most regard the mission statement as the cause and the vision statement to be the effect. Looking at this relationship differently, the mission statement is something to be accomplished, and the vision statement is the pursuit of that accomplishment.

DOI: 10.1201/9781003255116-4

Most small organizations do not have a written mission statement but rely on the implicit understanding of what is most important to the organization—what it stands for. Preparing a written mission statement enables the organization to state its purpose explicitly. In addition, the written mission statement should provide the reader with a summary of the organization's principles and culture. Internal and external stakeholders will be keen to know how the organization's values will determine how the mission is executed moving forward. Sustainability may be embedded into the operational objectives and the organization's values to become part of how the organization operates every day.

Many organizations seek to be a good neighbor and positive contributors to the community. Their work should not cause harm in the neighborhood. Organizations need members or employees to create the products and services associated with their operations and mission. If the community suffers from environmental, social, and economic problems, the organization will likely have trouble recruiting members and retaining them at this location. It is always easier to attract and retain members in an organization when a community grows sustainably.

Having a beneficial relationship with the local community is vital to securing and maintaining its "social license to operate." It is possible to aspire to values that will set the organization apart from its competitors by enabling mutually beneficial collaborations and partnerships with the community and other organizations. Sustainability and social responsibility allow the organization to build long-term relationships with local suppliers and customers while enhancing its reputation as a good neighbor in the community.

An organization can use its mission statement to communicate a good sense of what it is seeking to achieve in sustainability to its internal and external stakeholders. In this manner, the organization can share its legitimacy with the community and seek new members, employees, or trusted partners who identify with its stated purpose. Unfortunately, some organizations focus their mission statement on making a profit. This kind of statement may not be viewed favorably by some external stakeholders because it will seem that the organization places profits above addressing the stakeholders' interests. A sustainability program helps create responsible mission statements and objectives that demonstrate the value of its environmental stewardship, its focus on social well-being, and the shared value associated with these contributions. However, these contributions must not be seen as something provided separately by an organization that struggles with coming to terms with its place and role within the community.

ORGANIZATION'S OBJECTIVES

Once the mission statement is explicitly stated, linking the mission to the objectives is crucial.[1] This will help provide the means for accomplishing the mission. At the highest organizational level, strategic objectives are statements of broad intent strategically linked to the mission statement. Like the mission statement, these strategic objectives typically have a window of accomplishment of approximately 5–10 years. Ideally, an organization should have three to five strategic objectives derived from its mission statement. Each strategic objective should be concise, specific, and able to be understood by the stakeholders. It is considered good practice to state each strategic objective in a single sentence with fewer than 25 words. Many involved in organizational development think objectives to be "continuous," while the goals have a clear beginning and end associated with them. In this way, the objectives provide the bridge between an inspirational mission and the clearly stated goals that will provide a feedback loop from the bottom of the organization to prove that the strategic objectives are on target.

Objective setting does not end at the strategic level. Objectives must also be prepared for each of the organization's operating levels, local operations, and individual work units. At the operational level, objectives establish the "outcomes" sought through the processes and operations and explain what the organization aims to accomplish at these operating levels. Objective setting is a top-down process that must extend from the strategy-setting level to the lowest operating levels of the organization. The objectives at the lowest operating levels must be linked to objectives in the next higher level and be directly traceable to the strategic objectives. Objectives at the lowest levels of operations help motivate the members or employees to work effectively to help the organization realize its overarching strategic objectives. The operational objectives are specific and short-term to be used in the operation's day-to-day activities. The relationship between an organization's overarching strategic objectives and its operational objectives is a key factor in meeting its objectives in an uncertain world. Long-term organizational success is only likely to happen if short-term operational activities are consistent with the long-term strategic intentions. This is important to keep in mind when embedding the sustainability program into this objective structure. Sustainability objectives must be aligned with the operational objectives and directly support the strategic objectives as shown in Figure 4.1.

GOAL SETTING AND OPERATIONAL EXECUTION

An organization's strategy is implemented at the operational level by defining work goals and developing action plans that help people reach these goals. Goals provide a bottom-up feedback loop to test the top-down objectives. They state the specific, measurable results that specify how much or what will be accomplished by when. Goals linked to action plans let management know whether operations meet their immediate operational objectives. An action plan supports each work goal.[2] These action plans provide a prioritized set of activities that must be achieved to realize the work goal. Every action is subject to the question, how does this effort help the organization's operational and strategic objectives? This question allows the member or employee to consider how the organization operates with its processes and activities. No activity performed within the organization should be a stand-alone effort separate from the organization's main activities and processes.[3]

Every member of an organization, whether in a supporting position or an operational role at any level, should have one or two goals to work toward. There is evidence that suggests the chance of effectively achieving two or three goals is high; however, the more goals that someone tries to accomplish at one time, the more

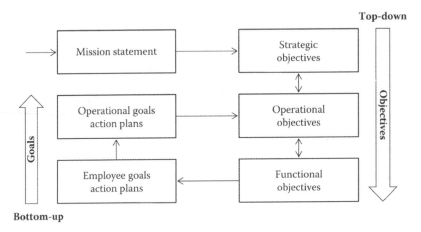

FIGURE 4.1 Relationship between objectives and goals in an organization.

likely it will be that none of the goals are attained.[4] The suggestion is to narrow the focus of each organization member to one or two "wildly important" activity goals in the short term and consistently invest the person's time and effort in achieving those goals. These wildly important goals will focus on the significant opportunities and threats that are prioritized by using risk assessment. It is not about selecting goals that provide sustainability results or meet other needs. Instead, it is about setting goals so fundamental to the organization's mission that achieving them ensures that the strategic objectives will be realized. Once the activity goals are selected, several lead and lag measures are used to track the progress following the action plan. The measures are carefully tracked to help the members or employees create a "cadence of accountability."[4]

Sustainability programs are often operated as a series of initiatives with their own goals, leading to a common and ineffective pattern of organizations setting objectives and goals as separate efforts or interchanging the two terms. Hence, there is no organized functioning of the organizational structure that is presented in this chapter. When considering that the organization is operating in an uncertain world, it is prudent to coordinate objectives and goals with the risk management efforts. The strategy should always align with the mission statement and the stakeholder interests. Organizations must create an efficacious strategy that can effectively address the mission statement through the auspices of its overarching strategic objectives.[5]

WRITING OR REVISING AN ORGANIZATION'S MISSION STATEMENT

Many larger organizations have written a mission statement. The mission statement comes from a parent organization at the top of a hierarchal structure in many cases. If there is an existing mission statement, it can be evaluated or revised before creating a set of strategic objectives. The mission statement process starts with a review or the drafting of the mission statement to make sure that it addresses four sets of questions:

- Who are we? What do we do? What are our products and services?
- Whom do we serve? Who finds these products and services of value?
- Why are we here? What value do we provide? What business problem, human need, or desire do our products and services fulfill? Have we identified the most critical values?
- What principles or beliefs guide the organization?
- There are many guides to writing mission statements on the Internet. Remember that a well-crafted mission statement:
  - Expresses the organization's purpose in a manner that inspires support and ongoing commitment to providing the organization with a unique identity
  - Is articulated in a way that is convincing and easy to understand
  - Uses proactive verbs to describe what the organization does
  - Is free of unnecessary jargon
  - Is short enough to enable stakeholders to recognize it easily

A well-written mission statement is an essential prerequisite for developing the strategic objectives.

DEVELOPING THE OBJECTIVES

To develop the strategic, high-level objectives, the drafting team needs to ask the following questions:

- What are the three to six areas in which the organization will continue to be actively involved over the next 5–10 years?
- What areas need attention to accomplish what is found in the mission statement?
- Do the objectives convert the mission statement into action?
- Do the objectives help to sustain the organization's competitive advantage?
- Is there a risk management program in place for prioritizing the opportunities and threats found in the organization's internal and external operating environment (context)?

The objectives of a parent organization affect the objectives throughout an organization. Business or divisional objectives are derived from the corporate objectives but are specific to these levels of the organization. Department objectives are derived from the divisional objectives to maintain the linkage back to the strategic objectives. This progression is repeated down to the individual member or employee level. It does not matter whether the organization is a business or a not-for-profit. Similarly, the time frame for objectives can range from five to ten years for long-term strategic objectives to one year for an objective at an individual level.[6] Every organization needs to have more than one objective at the strategic level. Objectives are always derived from the mission statement. A sustainability program can set strategic objectives only if there is a focus on sustainability embedded within the operational or functional levels, which support the strategic objectives at the highest level in the organization.

The mission and the strategic objectives should not be unchangeable.[6] Mission statements form the foundation for objectives, and both can change over time. In addition, factors in the external operating environment can create opportunities and threats as the uncertainty of the operating environment is altered.

DEVELOPING GOALS AND ACTION PLANS

Ideally, both supervisors and the members or employees at the lowest operational levels should define the goals. The supervisor helps clarify the operational objectives and their role in setting the strategic direction. The employees should work as a team to develop the goals to express their commitment to the strategic direction. Achieving all the individual goals in an area should make it possible to achieve the operational objective that has been assigned to that supervisor. Each member or employee would have one or two clearly stated and measurable goals.

Several essential questions must be addressed within this goal-setting process:[4]

- What element of the performance of our work can best help achieve the operational goals?
- What are our greatest strengths that can be leveraged to achieve the goal?
- What area of our past performance needs to be improved to ensure the goal is achieved?

The focus of these questions should be on defining the goals. It is critical to identify the goals that promise the most significant positive consequence for the operational objective that supports the strategic objectives. Therefore, the answers to these questions should be ranked by effect—for the organization as a whole, not for the area's performance.

Once there are a few candidates, the team tests the goal against the so-called SMART elements:

- Specific
- Measurable
- Achievable
- Realistic
- Timely

Once these elements are satisfied, the worker or team can assist the supervisor in defining the goal so that an action plan can be prepared.

An action plan provides the means to ensure that the achievement of the goal will help the organization meet its objective. The action plan needs to be complete, precise, and current. It should include information and ideas gathered during the planning of the goals. The following information is needed:[7]

- Describe the goals to be achieved.
- Describe the specific, measurable, and attainable outcome-based goals.
- What tasks, actions, or activities will accomplish the goal?
- Who will conduct and be held responsible for these actions?
- When will the activity take place? Provide an appropriate timeline.
- Secure the allocation of the resources (e.g., money and people) needed to conduct the activity.
- How will the activity, along with its successes and opportunities for improvement, be communicated within the organization?

If the organization does not have a formal action planning process, there are some excellent examples to follow.[2, 7] Management should review the action plans regularly as part of their effort to ensure that the organization achieves its objectives.

If the sustainability program is embedded within the organization, the operational objectives will require sustainability goals to be established as described above. Alternatively, the sustainability goals can be accommodated as part of the organization's other structural elements, as described in the following chapters.

UNCERTAINTY AFFECTS THE OBJECTIVES AND GOALS

The "top-down" objectives and "bottom-up" goals guide the system of processes and activities to help the organization meet its objectives for sustainable success. An organization's operations are the day-to-day activities responsible for delivering the strategy. However, some events and decisions happen along the way, creating opportunities and threats for the operations. These opportunities and threats are identified by characterizing the organization's context at the operational level.

ESSENTIAL QUESTIONS FOR ORGANIZATIONAL OBJECTIVES AND LEADERSHIP

1. Why are sustainability goals established without regard for the organization's strategic objectives and goals?
2. How can the sustainability goals influence the organization's strategic objectives? How can the sustainability initiatives be defined around the employees' goals?
3. Why are sustainability goals focused on producing results rather than helping the organization meet its strategic objectives?
4. How can the stakeholders be affected when the organization meets its strategic objectives?
5. What happens when the results of the employees meeting their goals do not meet the operational objectives at that level?

REFERENCES

1. Geneva, ISO (International Organization for Standardization), *Quality Management Systems: Fundamentals and Vocabulary*, Geneva: ISO, 2014.
2. R. Pojasek, "Implementing your pollution prevention alternatives," *Pollution. Prevention Review*, vol. 7, 83–88, 1997.
3. ISO (International Organization for Standardization), *Risk Management: Principles and guidelines*, Geneva, 2009.
4. C. C. S. A. H. J. McChesney, *The 4 Disciplines of Execution*, London: Simon & Schuster, 2012.
5. P. Hopkin, *Fundamentals of Risk Management: Understanding, Evaluating and Implementing Effective Risk Management*, Kogan London: Page Publishers, 2012.
6. R. Barnat, *Strategic Management: Formulation and Implementation*, 2. Retrieved August 3, Ed., 2014.
7. Community Tool Box., *Developing an Action Plan.*, Retrieved August 3, 2015, from http://ctb.ku.edu/en/table-of-contents/structure/strategic-planning/   develop-action-plans/main

5 Organization's Internal and External Context

ABSTRACT

Management of an organization involves many coordinated activities to control its activities, processes, operations, products, and services as it pursues its strategic objectives.

Risk management is a component of organizational management involving coordinated activities associated with uncertainty on those objectives. Effective risk management is fundamental to the organization's success, as it focuses on its performance against the objectives.[1] An organization's performance in relation to society and the economy and its impact on the environment has become a critical part of measuring its overall performance and its ability to continue operating effectively. Because risk is the effect of uncertainty on achieving strategic objectives, it is crucial to examine the influences and factors in the internal and external context that can affect the organization's objectives.

CONTEXT OF THE ORGANIZATION

Organizations operate in an uncertain world. Whenever an organization seeks to meet its objectives, there is a chance that everything will not go according to plan. Moreover, it is possible that the organization will not achieve its objectives even if the objectives were carefully planned. What lies between the organization and its objectives is uncertainty. Uncertainty represents a deficiency of information that leads to an incomplete understanding of what can happen that would threaten or enhance the organization's ability to meet its objectives.

The operating environment within which the organization operates includes many sources of uncertainty. This uncertainty creates "effects." These effects can be negative or positive. Adverse effects are referred to as "threats." Positive effects are referred to as "opportunities." These effects create positive or negative consequences[2] that affect the organization's ability to meet its strategic objectives. Consequences are the outcomes of an event or decision affecting the organization's objectives. Effective risk management helps the organization understand its opportunities and threats and address them as appropriate to manage the consequences and thereby maximize its chance of achieving its objectives by managing the uncertainty. It is important to remember that risk is associated with the performance of the organization's strategic objectives.

INTERNAL CONTEXT

An organization's internal context is the operational environment it seeks to achieve its objectives. The internal context includes anything that the organization has

DOI: 10.1201/9781003255116-5

control over or has a sphere of influence. The sphere of influence is an important concept. It represents the range and extent of political, contractual, economic, or other relationships through which an organization can affect the decisions or activities of individuals or organizations.[3] The internal context can include the following:[2]

- Governance, organizational structure (including relationships with a parent organization), roles, and accountabilities
- Policies, objectives, and strategy
- Operational capabilities understood in terms of processes, operations, and resources
- Decision-making, knowledge, and sense-making
- Engagement with the internal stakeholders
- Relationships with other organizations within its value chain
- Organization's culture
- Standards, guidelines, and operating models adopted by the organization
- Honoring contractual relationships

An organization must consider everything internal and relevant to its mission, strategic objectives, strategic direction, processes, and operations. It needs to understand the influence these considerations could have on its sustainability program and its intended results.

EXTERNAL CONTEXT

The external context includes the broad external operating environment in which the organization operates. Influences and factors in the external context can include the following:[2]

- Cultural, social, political, legal, regulatory, financial, technological, economic, natural, and competitive
- Key drivers and trends exerting positive and negative consequences that affect the objectives
- Relationships with external stakeholders, along with their perceptions, values, and interests

Organizations must assess their external operating environment to determine and characterize the crucial influences and factors that might support or impair their ability to manage the identified opportunities and threats. These factors are identified by examining the conditions, entities, and events that determine the associated opportunities and threats that may influence the organization's activities and decisions.[4]

Because the external operating environment affects the operations, people involved in supporting the operations are interested in the information created in the characterization of the external environment. The following elements are often considered:[5]

- Material resources: suppliers, real estate, and brokers
- Human resources: labor market, schools, and unions
- Financial resources: banks, investors, granting agencies, and contributions
- Markets for products and services: customers, clients, and users
- Competitive environment: competitors and competitiveness
- Technology: hardware, software, information technology, and production techniques
- Economic conditions: inflation rate, signs of recession, rate of investment, and growth
- Government oversight: regulations, taxes, services, the court system, and political process
- Sociocultural: demographics, values, beliefs, education, religion, work ethic, and the green movement
- International: exchange rate, competition, selling overseas, and regulations

An organization needs to consider its relationship to each of these elements in terms of its strengths, weaknesses, opportunities, and threats.[4]

SCANNING THE ORGANIZATION'S OPERATING ENVIRONMENT

Organizations are connected to an outside world with a constantly changing set of influences and factors. There is often similar change going on within the organization. Organizations should monitor the context to identify, assess, and manage the opportunities and threats associated with change. If the organization can analyze these changes that can affect its performance, it can begin making decisions and developing a strategy for operating in this uncertain world.

Understanding the organization's context provides insight into the internal and external influences and factors that impact the way organizations operate at the community level and influence the decisions that will be made by leadership at that location.

SCANNING THE INTERNAL ENVIRONMENT

Understanding the internal context prepares the organization for managing the opportunities and threats originating within its processes and operations. The internal context is also vital to implementing any risk management activity by the organization. Most factors associated with the internal context are within the organization's control or sphere of influence.

A methodology widely used in the project management field provides a set of "influences" helpful in scanning the internal operating environment. This TECOP tool contains the following influences:[6]

- Technical: Information and communications technology (hardware and software), internal infrastructure (including its condition), knowledge sharing, research and development, assets, required skill levels and competency of workers, and innovation efforts

- Economic: Financial management, financing, cash flow, return on investment, capital reserves, taxes, royalty payments, insurance, and the commercial viability of the organization
- Cultural: Demographics, and collective attitudes and behavior characteristics of the organization
- Organizational: Capabilities, policies, standards, guidelines, strategies, management systems, structures, and objectives
- Political: Governance, internal politics, decision-making systems, stakeholders, roles, and accountabilities

There is another version of TECOP that replaces *culture* with *commercial*. However, to determine the internal context, the cultural category is of great importance for understanding the influence of people on an organization's operations.

When considering the processes and operations of the organization, a SWIFT (i.e., "structured what-ifs technique") tool[2] can be used to supplement the characterization of the internal context using the TECOP analysis. SWIFT is a method that enables a facilitator to guide a team through a systematic evaluation using "prompts" to determine how the processes and operations could be affected by opportunities and threats found through the TECOP analysis. Deviations from processes and operations create additional opportunities and threats that must be considered internally.

SCANNING THE EXTERNAL ENVIRONMENT

The external context is characterized by conducting a broad scan of the external operating environment. A PESTLE (also called PESTEL) analysis tool systematically assesses the influences and factors that create opportunities and threats in the external operating environment that the organization does not control. The influences associated with the PESTLE analysis include the following:[7]

- Political: Factors include the extent to which governments or political influences are likely to impact or drive global, regional, national, local, and community trends or cultures.
- Economic: Factors include global, national, and local trends and drivers; financial markets; credit cycles; economic growth; interest rates; exchange rates; inflation rates; and the cost of capital.
- Societal: Factors include culture, health consciousness, demographics, education, population growth, career attitudes, and an emphasis on safety.
- Technological: Factors include computing, technology advances or limitations, robotics, automation, technology incentives, the rate of technological change, and research and development.
- Legal: Factors include legislative or regulatory issues and sensitivities.
- Environmental: Factors include global, regional, and local climate; adverse weather; natural hazards; hazardous waste; and related trends.

Gathering the information to prepare the external context can be time-consuming and difficult. Having an independent analysis conducted by a local team under contract could facilitate the ability to determine the opportunities and threats associ-ated with each of the factors. However, the people working within the organization may have some unique perspectives that a contractor might not develop during the life of the contracted effort. Once there has been some meaningful engagement of the internal stakeholders, they can use their knowledge of the external operating environment to complement this effort.

ADAPTING TO A CHANGING EXTERNAL CONTEXT

Uncertainty in the external context may create a need to change the organizational structure and operating behaviors. An organization in a specific external operating environment is managed and controlled differently from operating in an uncertain external context. Therefore, organizations need to have the correct fit between inter-nal structure and the external environment.

Uncertainty can be expressed in terms of volatility, uncertainty, complexity, and ambiguity:[8]

- Volatility: Unexpected or unstable challenge that may be of unknown dura-tion, but knowledge about it is often known and available. How volatile is the current situation? What are the nature and dynamics of change, and how can the change affect the organization? What is the nature and speed of the forces of change?
- Uncertainty: Despite a lack of other information, an event's basic cause and effect are known; change is possible but not a given. How much predict-ability is expected, and which areas of the organization's structure have the least levels of certainty? What factors should be analyzed around the lack of predictability, the prospects for surprise, and the sense of awareness and understanding of factors and events?
- Complexity: A situation has many interconnected parts and variables; some information is available or can be predicted, but the volume or nature of that information can be overwhelming to process. How complex are the context, the operating model, and the environment within which the organization operates? What is the combination of forces, the confounding of factors, and the chaos and confusion found in the external context?
- Ambiguity: Causal relationships are completely unclear; no precedents exist, facing "unknown unknowns." What level of ambiguity is the orga-nization currently facing, and what is expected in the future? In what areas is the organization facing causal relationships, and how are they likely to affect it? What are the key factors around the haziness of reality, potential for misreading, or mixed meaning of conditions and cause and effect?

Volatility, uncertainty, complexity, and ambiguity are components of complexity theory or the science of complexity. There are a growing number of studies to apply practical approaches to organizations so they can strategize and change.[9]

FINDING OPPORTUNITIES AND THREATS

Scanning involves the analysis of events, trends, and relationships in the organization's internal and external operating environment. The information obtained from such an effort can assist organization leaders in planning a strategy that can guide the organization in the future. It is crucial to scan the external operating environment to understand the forces of change, such as volatility, uncertainty, complexity, and ambiguity. Scanning involves looking at information (i.e., viewing) and looking for information (i.e., searching).[10]

SCANNING METHODS

One method to help understand the external operating environment is to review a wide range of different sources:[1]

- Laws and regulations
- Newspaper
- Electronic media
- Newsletters, magazines, journals, and books
- Reports and presentations
- Interviews

All the information from the strategic planning process should be reviewed. Data from the organization's documented information is helpful for both the operations and the supply chain.

Once information from all sources has been reviewed, it is time to start using the PESTLE analysis method. The acronym letters refer to the influences in the external operating environment. Each influence has several factors. It does not matter how the factors are allocated to the different influences. The scanning activity aims to identify as many factors as possible.[11]

To find out which factors can yield valid opportunities and threats, it is important to have the scanning team learn how to ask effective questions. Questioning is a vital tool that helps obtain data, knowledge, and wisdom. However, the information obtained is only as good as the questions asked. Questions contribute to the success or failure of finding opportunities and threats as the team seeks to understand a complex and changing external context.[12]

EFFECTIVE QUESTIONING

The scanning team needs to have a plan to use the questions to catalyze insight, innovation, and action on the part of the organization. Each of the following steps is of critical importance to the discovery of opportunities and threats:[12]

- Assess the current situation with the influence–factor combination.
- Discover the big questions that can help identify and clarify the opportunities and threats.

- Create images of possibilities and scenarios of how the organization may be affected.
- Evolve and create strategic opportunities and threats.

Many organizations use scanning for marketing as well as for strategy. Keep in mind that the methods and outcomes are quite different. Sustainability practitioners are usually keen on finding opportunities in both the internal and external context. Therefore, it is important to engage the sustainability team in the scanning effort. If there is no sustainability focus in place, the scanning team needs to reach out to people involved in the stakeholder engagement process. It is likely that the external stakeholders are already savvy about what is happening in the neighborhood and the community. Exchanging information with the stakeholders will help improve the effectiveness of the engagement process.

The scanning team should consider the following methods to help improve the effectiveness of their effort:[12]

- Engage in shared conversation.
- Convene and host "learning conversations."
- Include diverse perspectives in all conversations.
- Foster shared meaning.
- Nurture communities of practice.
- Use collaborative technologies.

All these are essential elements of effective questioning in a learning organization.

CRITICAL THINKING

After the questioning, the focus shifts to critical thinking; by its nature, critical thinking demands recognition that all questioning:[13]

- Stems from the point of view and occurs within a frame of reference
- Proceeds for some purpose—to answer a question
- Relies on concepts and ideas that rest on assumptions
- Has an informational base that must be interpreted
- Draws on basic inferences to make conclusions that have implications and consequences

Each of these dimensions of reasoning is linked to the others. Failure to recognize the links in any of these steps will impact the other steps. Therefore, the critical thinking process needs to be carefully monitored.[13]

Critical thinking is the practice of analyzing and evaluating thinking to improve the information derived from this thinking. Critical thinking is the purposeful, reflective, reasonable, and self-regulatory process of thinking out possible opportunities and threats from each of the influences and their factors and determining the evidential, conceptual, methodological, and contextual considerations upon which judgment on these opportunities and threats is based.[13]

The critical thinker:[13]

- Raises questions and discusses potential opportunities and threats associated with every influence–factor, being careful to formulate them clearly and precisely
- Gathers and assesses relevant information using critical thinking to interpret the information effectively
- Thinks open-mindedly within the critical thinking realm to recognize and assess assumptions, as well as the implications and consequences from various interpretations and inferences
- Communicates effectively with others to determine the opportunities and threats associated with each influence–factor combination

Critical thinking is analytical, judgmental, and selective. This contrasts with creative thinking, which is more adaptive to a changing context, especially when escaping from a preordained pattern. Thinking creatively involves generating ideas that are often considered unique and plausibly effective. To find the right opportunities and threats for each influence–factor combination, it is necessary to have some creative thinking capability present on the scanning team or to find that information via a literature search or interviews. Critical thinking will remain the dominant force.[13]

STABILITY OF THE EXTERNAL CONTEXT

Researchers have noted that a stable external context creates an internal context with a mechanistic bent favoring standard rule, procedures, a clear hierarchy of authority, formalization, and centralization. This contrasts with an external context in rapid change, where the internal context appears to be much looser, free-flowing, and adaptive. Rapid change spawns a flexible hierarchy of authority and a decentralized decision-making process. Organization members or employees are encouraged to deal with opportunities and threats by working directly with one another, using collaborative teams, and taking an informal approach to assigning tasks and responsibilities. This helps the organization adapt to continual and sudden changes in the external environment.[5] Organizations that are most successful in uncertain external contexts keep everyone in close touch with influences, factors, opportunities, and threats in the external context. This can hasten the response that may be necessary.

SUSTAINABILITY AND CONTEXT

Organizations operating without a focus on sustainability seek to control internal threats primarily. In a financial environment, this is referred to as "internal controls." In an internal context, this is referred to as "operational controls." Scanning the external operating environment often involves those interested and involved in the organization's sustainability efforts. These sustainability efforts focus on effective process and efficient operations within the internal context and on searching for opportunities and threats in the external operating environment through their involvement with the stakeholder engagement program. Without focusing on sustainability, the stakeholders are regarded as "interested parties."

Sustainable organizations with a high level of stakeholder engagement tend to be more vigilant about changes in the external context.

In some organizations, the information about the internal context can be determined by reviewing existing practices, plans, processes, and procedures. However, many organizations operating at the community level tend to operate informally. Context review team members need to ask questions of the members or employees to determine how they function within the organization. The organization's leader could access the ability of the organization to achieve the identified objectives through some basic level of strategic planning (e.g., strengths, weaknesses, opportunities, and threats). It always helps to see how the organization has dealt with previous failures, incidents, accidents, and emergencies to obtain a complete picture of an informal internal context.

Changes in either the internal or external context create uncertainty for the organization. Organizations must manage the effects of uncertainty (i.e., the opportunities and threats) to lower the risk to meet their objectives. An increased level of uncertainty is often associated with the inability of internal decision-makers to obtain sufficient information about the context factor and the opportunities and threats associated with these factors. Uncertainty increases the risk of meeting the objectives. Opportunities can contribute to the chances that the objectives will be met. Threats are likely to lead to changes that the objectives will not be met.

A complex external operating environment is one in which the organization interacts or is influenced by many diverse PESTLE influences and factors. In contrast, a simple external operating environment leads to an organization that only interacts with or is influenced by a few similar factors as determined using the PESTLE analysis.[13] There is also a stable–unstable dimension associated with the external context. In a stable external operating environment, the PESTLE influences and factors remain essentially the same over months or years. However, during unstable conditions, the PESTLE influences and factors change rapidly. Therefore, it is imperative to constantly monitor the external context under unstable conditions.[13] This is a role that the sustainability team is particularly suited for.

A PATH FORWARD

The context analysis team might consider the following "looking forward" actions:

- Reviewing or creating the risk management and sustainability policies
- Analyzing the organization's resources (e.g., financial capital, human capital, assets, and materials productivity) and knowledge management capability
- Performing a stress test of the decision-making processes, including sensemaking and knowledge management

- Creating a cohesive set of organizational systems of management for the use of a plan–do–check–act sequence of processes
- Enhancing the organization's contractual and informal relationships that involve adherence to a "supplier code of conduct" as a condition of collaborating at any level with another organization

Organizations will use these actions to protect the operations associated with the internal context and otherwise manage the level of uncertainty to help meet their objectives. To thrive in a world characterized by uncertainty, an organization needs to turn to risk management and sustainability to help take advantage of the opportunities and offset the threats. However, treating threats to make them less unacceptable always leads to additional threats when operating in a system. Therefore, sustainability efforts need to be continuously vigilant concerning opportunities and threats associated with a changing external context. These efforts will also help engage the external stakeholders to seek their support in this effort.

ESSENTIAL QUESTIONS FOR ORGANIZATION'S INTERNAL AND EXTERNAL CONTEXT

1. How do changes that occur in the internal and external operating environments affect the ability of the organization to meet its strategic objectives?
2. How is sustainability affected by changes in the internal and external context? How does the corporate sustainability manager consider this when preparing the sustainability report?
3. Why is it important for a corporation to understand that all its facilities (i.e., organizations) are "different" because of the context, even if they have identical products and services?
4. How can the PESTLE and TECOP tools help standardize the scanning processes at different facilities to maintain the corporate sustainability program?

REFERENCES

1. H. 436, *Risk Management Guidelines, AS/NZS (Standards of Australia/Standards of New Zealand)*, Sydney: SAI Global Press, 2013.
2. Geneva, *Risk Management: Principles and Guidelines. ISO 31000*, Geneva: ISO (International Organization for Standardization), 2009.
3. Geneva, *Social Responsibility Guidance*, Geneva: ISO (International Organization for Standardization), 2010.
4. H. 203, *Environmental Risk Management, AS/NZS (Standards of Australia/Standards of New Zealand)*, Sydney: SAI Global Press, 2000.
5. R. Daft, *Organization Theory & Design.*, Mason, OH: South-Western Learning, Centage Learning, 2013.
6. J. Talbot, "Organizational risk management and sustainability," *(Internal Context. Blog)*, vol. 2. Retrieved June 29, Ed., 2011.

7. J. Talbot, "The external context: What's outside the door?," *Blog*, vol. 2. Retrieved June 29, Ed., 2011.

8. N. a. L. G. Bennett, "What VUCA really means to you," *Harvard Business*, vol. 9. (. 2. Review, Ed., 2014.

9. O. Serrat, "Understanding complexity," *Knowledge Solutions*, vol. 2. Retrieved June 29, Ed., Manila: Asian Development Bank, 2009.

10. W. Choo, "Environmental scanning as information seeking and organizational learning," *Information Research*, vol. 7, no. 1, 2001. [Online]. Available: https://www.informationr.net/ir/7-1/paper112.html

11. F. (. M. Ebooks), "PESTLE analysis: Strategy skills," Vols. Retrieved January 31, 2016, 2013.

12. O. Serrat, "Asking effective questions. Knowledge solutions," Vols. Retrieved June 29, 2015, 2009.

13. O. Serrat, "Critical thinking. Knowledge solutions," Vols. Retrieved June 29, 2015, 2011.

6 Engagement with Stakeholders and Social License to Operate

ABSTRACT

Every organization comes face-to-face with a wide variety of different kinds of stakeholders as it begins to develop sustainability by characterizing its internal and external operating environment. A stakeholder is an interested party in an organization's decisions or activities. Interest refers to the actual or potential basis of a claim or demand for something that is owed or to demanding respect for a right. The claim is not necessarily a financial demand but a claim to be heard. This interest gives them a "stake" in the organization. However, this relationship is usually not formal or even acknowledged by the stakeholder or the organization. Stakeholders are also referred to as "interested parties." The relevance or significance of interest is determined by the principles of risk management and sustainability. Organizational sustainability is the willingness of an organization to operate in a transparent and accountable manner and compliance with all applicable laws and regulations. It is expected that the organization practices its responsibilities with the stakeholders' interests in mind within its activities, processes, and operations.

STAKEHOLDERS

Identification of and engagement with stakeholders are fundamental to the practice of sustainability by an organization. Every organization should determine who is interested in its activities, processes, decisions, products, and services. It needs to determine the consequences of its activities when conducting the internal and external operating environment scanning exercises. While an organization is usually clear about the interests of the owners, members, customers, or other constituents of the organization, there are often people and organizations outside of the control of an organization that have rights, claims, or other specific interests that need to be addressed that are not being addressed. Not all the stakeholders belong to organized groups that have the purpose of representing their interests. In some cases, the "interest" exists whether or not the parties are aware of the mutual interest.

Many organizations are unaware of all their stakeholders, and stakeholders may not be aware of the potential of an organization to affect their interests. To address these conditions, an organization should:[1]

- Create a means of identifying its stakeholders and keep that list current
- Recognize and have due regard for the interests and other legal rights of its stakeholders and engage with them regarding their interests

DOI: 10.1201/9781003255116-6

- Assess and take into account the ability of stakeholders to contact, engage with, and influence the organization
- Take into account the relation of its stakeholders' interests to the organization's risk management and sustainability programs, as well as the nature of the stakeholders' relationship with the organization
- Consider the views of stakeholders whose interests are likely to be affected by an activity, process, or decision even if they have no formal role in the governance of the organization or are unaware of these interests

An organization needs to understand the relationship between the various stakeholders' interests that are affected by the organization while also considering the interests of society as a whole—starting with the community. Although stakeholders are a segment of the larger society, they may have an interest that is not consistent with society's expectations.[1]

An effective means for an organization to identify its sustainability responsibilities is to become familiar with the practice of scanning its internal and external operating environments while realizing that the stakeholders are the "face" that can be put on the opportunities and threats that have been identified.

INTERNAL STAKEHOLDERS

Some stakeholders are involved within the organization, including members, employees, leaders, and owners. These stakeholders are interested in the organization's mission and strategic objectives. However, this does not mean that all their interests will be the same. Competent, empowered, and engaged people at all levels in the organization are critical to the organization's success in meeting its objectives in an uncertain world. Recognition, empowerment, and enhancement of competence help facilitate people's engagement in achieving effective processes and efficient operations in the organization.

It is important to remember that internal stakeholders are also connected to external stakeholders through their family and friends in the community. Since everyone belongs to multiple organizations, it is essential to have the members or employees assist with monitoring the external operating environment and identify which key external stakeholders are known to them.

Another point to remember is that the parent organization is linked to all its organizations through the internal context and reporting structure. This provides some consistency throughout a corporation, but the individual facilities are still different because of context.

EXTERNAL STAKEHOLDERS

All the organizations that are identified in the scan of the external operating environment have people that become external stakeholders. It is crucial to begin searching for external stakeholders with these people and their organizations. The organization and its parent organization may have a sphere of influence on some of the external stakeholders.

Some organizations consider their value chain organizations as internal stakeholders since contractual arrangements or other factors often create a sphere of influence. However, the power level is reduced in the second and third tiers of the value chain.

IDENTIFICATION OF STAKEHOLDERS

An organization should determine what other organizations or individuals are interested in its activities, processes, decisions, products, and services. While this is often accomplished as a component of scanning the internal and external operating environment, extra effort is warranted to enable the organization to understand the relevance of its opportunities and threats and the consequences of these effects of uncertainty on the stakeholders. Putting a "face" on these opportunities and threats helps the organization when it conducts the uncertainty analysis. However, the identification of stakeholders does not replace the considerations of the broader reach of society in determining norms and expectations of the organization's activities and decisions.

Parent organizations (e.g., corporations) will have global stakeholders, such as the influential nongovernmental organizations (NGOs), and multilateral groups, such as the United Nations and the International Labour Organization. Organizations at the community level will deal with local government officials, community government departments, local community board members (e.g., zoning board), neighbors, local advocates of various causes, and people involved in community service. Many people in a community have an "interest" in what an organization is seeking to accomplish with its strategic objectives.

To identify stakeholders beyond its scan of the external operating environment, an organization should consider the following questions:[1]

- To what organizations are there legal obligations?
- Who might be positively or negatively affected by activities, processes, decisions, products, and services?
- Who might express concerns about these activities, processes, decisions, products, and services?
- What involvement have other organizations or individuals had in the past when similar concerns needed to be addressed?
- Typically, who helps the stakeholder organization address specific consequences?
- Which stakeholders can affect the ability of an organization to meet its responsibilities?
- Who would be disadvantaged if excluded from engagement?
- Who in the community or value chain is affected?

Some sustainability practitioners refer to interest as an "issue." An issue is defined[2] as "a point or matter in question or dispute, or a point or matter that is not settled and is under discussion or over which there are opposing views or disagreements." When stakeholders have issues rather than interests, the organization has waited too long to engage with them. Learning about an issue signifies that an impasse has been reached and that mediation may be necessary to get past this impasse. An organization could jeopardize its social license to operate by not choosing to deal with stakeholder issues.

To be sure that an organization has satisfactorily dealt with the identification of its stakeholders, it should consider each of the following tasks:[1]

- Have a process for identifying both its internal and external stakeholders while considering the process for conducting the internal and external operating environment scans
- Recognize and have due regard for the interests, as well as the legal rights, of its stakeholders and engage with them to better understand their interests
- Recognize that some stakeholders can significantly affect the level of uncertainty of an organization and keep it from achieving its strategic objectives
- Assess and consider the relative ability of stakeholders to contact, engage with, and influence the organization
- Consider the relation of its stakeholders' interests to the broadest expectations of society and sustainability, as well as the nature of the stakeholders' relationship with the organization
- Consider the views of stakeholders whose interests are likely to be affected by a decision or activity—this is a perspective view that comes from an effective stakeholder engagement process

Stakeholder identification and engagement are critically important to address an organization's three sustainability responsibilities—environmental stewardship, social well-being, and the sharing of value between organizations in the community.

STAKEHOLDER ENGAGEMENT

Stakeholder engagement is defined[1] as "activity undertaken to create dialogue opportunities between an organization and one or more of its stakeholders, to provide an informed basis for the organization's decisions." This term does not specifically address what constitutes engagement. Some quality management texts have defined[3] engagement for members or employees as "the level of connection felt with their employer, as demonstrated by their willingness and ability to help the organization succeed largely by providing discretionary effort on a sustained basis." In this case, engagement refers to the emotional and intellectual commitment to accomplishing the organization's strategic objectives. Employees feel engaged when they find personal meaning and motivation in their daily efforts and receive positive feedback and interpersonal support. An engaged workforce is made possible through developing trusting relationships, good communication, a safe work environment, empowerment, and performance accountability. A job is only a job until the worker identifies with it and shares its meaning. Employees must know how they contribute to their organization's success and the achievement of the strategic objectives.[3]

External to the organization, engagement can involve similar activities:

- Face-to-face meetings with active listening on both sides
- Involvement taking place over time
- Seeking to monitor the effectiveness of the engagement process

Engagement should be expressed as positive, proactive involvement.[4] When engaged with an organization, individuals must be involved. Interactions occur, information is exchanged, and operational friction is reduced. As with employees, effective engagement with external stakeholders also involves a strong emotional bond to the organization. Unfortunately, few stakeholder engagements have ever reached this level. Some organizations still use surveys to gather information from internal and external stakeholders. However, surveys do not meet any of the three elements of engagement mentioned above.

Digital media is changing the way we engage with stakeholders. Being digital enables an organization to evaluate how its engagement with external stakeholders may present opportunities and threats. This means understanding how stakeholder behaviors and interests are developing outside the business, which is crucial to getting ahead of trends that can deliver or destroy value.[5] Digital tools also contribute to rethinking how to use new capabilities to improve stakeholders' engagement. Data and metrics can focus on delivering insights about stakeholders that, in turn, drive the ability to understand their interests. Digital tools enable an organization to constantly engage its stakeholders as part of a cyclical dynamic where internal processes and changes in the external operating environment continually evolve based on direct inputs from the stakeholders, fostering ongoing engagement and empowerment. Being digital is about using data to make better and faster decisions and moving those decisions lower in the organization. The use of digital tools also speeds the engagement process, enabling fast-moving, stakeholder-facing interactions over time.[5] Many small organizations are beginning to use digital media in their marketing efforts. They will learn how to extend it to stakeholder engagement.

Another means for improving stakeholder engagement is borrowed from the more traditional voice of the customer (VOC) method.[4] A VOC program is a "channel for acquiring business insight about customers and what is important to them." Capitalizing on customer feedback requires a strategic and ongoing dedication to hearing, active listening, understanding, and acting on the customer's "voice: through a formal program built on:[4]

- Active listening: A mechanism that allows all customers to share their compliments and comments about their experiences with an organization's product or service.
- Pulse monitoring: A recurring and systematic means of tracking changes in business outcomes, their leading indicators, and the influential drivers by periodically contacting and talking to a statistically representative sample of customers. This is now done digitally instead of using polling.

Listening to customers is now starting to include the remainder of the external stakeholders. This needs to be a central element in the scanning of the external environment as well. Many organizations are beginning to use these VOC techniques with members or employees.[4]

PUTTING ENGAGEMENT INTO PRACTICE

Some actions for engaging members or employees of the organization are[6]

- Recognize that sustainability is embedded in the work that they do.
- Promote collaboration throughout the organization to ensure it meets its three sustainability responsibilities.
- Communicate face-to-face with people to promote understanding of the importance of their contributions to sustainability and to assess their satisfaction with the working environment that enables them to realize their commitments.
- Facilitate open discussion and sharing of knowledge and experience between members or employees.
- Recognize and acknowledge people's contribution and learning and the improvement that comes from their efforts.
- Enable their performance self-evaluation against their objectives for sustainability within their control.
- Communicate the results of sustainability and recognize all significant contributions.

When an employee is engaged with an organization, operational friction is reduced while an emotional bond is created. However, employee engagement must be self-motivated rather than simply participating in random sustainability initiatives.

A VOC-type program is much more than the use of an occasional survey. It recognizes the importance of customers (and employees) to an organization's success and a commitment to include stakeholder perspectives in decisions being made in every part of the operations. Essential questions for addressing sustainability as a critical component for stakeholder engagement and maintaining the organization's social license to operate can be found at the end of the chapter.

It begins with the opportunities and threats gathered in the internal and external context scans, continues with the practical and insightful analysis of stakeholder feedback (i.e., interests), and concludes with the creation of activities guided by the analysis to demonstrably impact how the organization engages all its stakeholders.[4] This is the way to embed stakeholder engagement into the organization's culture.

Within a sustainability program, employees are engaged, informed, incentivized, motivated, and rewarded for their contributions to the organization's sustainability program. This effort is often operated as a "behavioral change" program with various tools, techniques, frameworks, and approaches. When an organization commits to a VOC-style program, the bolt-on sustainability program has much more difficulty showing how it can add value through its contribution to stakeholder engagement. Sustainability needs to be built into the program.

SOCIAL LICENSE TO OPERATE

The social license to operate is defined as existing when an organization has ongoing approval within a community and from other internal and external stakeholders. This can also be expressed as having broad social acceptance or ongoing acceptance.[7] Social license to operate is rooted in the external stakeholders' beliefs, perceptions, and opinions. While the "license" is intangible to a great extent, an organization recognizes the importance of earning the acceptance or the highest level of approval from its stakeholders. This represents the greater community accepting the organization and its projects into their collective identity.[7]

The stakeholders in the community usually grant a social license to operate. Similar organizations might have a social license in one application and not have it for a similar application in the same area or neighborhood. Generally, the more expansive the environmental, social, and economic impacts of an organization and its activities, products, and services, the more difficult it becomes to get a social license.

The critical components of the social license to operate are:[7]

- Legitimacy: This is based on the norms of the community, which may be legal, social, and cultural, as well as formal or informal, in nature. Organizations must clearly understand the community's "rules of the game." Failure to do so risks rejection. In practice, the social license to operate comes from a stakeholder engagement program. Nothing should be left to chance. There must be a transparent exchange of information about the activities, decisions, products, or services.
- Credibility: The capacity to be credible is created by consistently providing accurate and precise information and complying with all commitments made to the community in the stakeholder engagement process. Documenting the understandings helps manage expectations and reduces the risk of losing credibility by being perceived as creating a breach of promises. However, many experts offer the advice of not making a verbal commitment since there are always areas open to reinterpretation at a later date in the absence of a permanent record.
- Trust: Trust, or the willingness to be vulnerable to the actions of another, creates a very high-quality relationship and one that takes both time and effort to create. Trust comes from shared experiences gained in the stakeholder engagement process. The challenge for the organization is to go beyond a traditional stakeholder engagement process and develop opportunities to collaboratively work together and generate the shared value within which trust can prosper.

Gaining and maintaining a social license to operate is complex. Difficulties arise when organizations are unable or unwilling to make the nominal investment of time to make things work well.[7]

There are several common problems associated with obtaining a social license to operate:[7]

- An organization sees the social license to operate as a series of tasks, while the stakeholders and community grant the license based on the quality of the relationship. As a result, a cultural mismatch often fails.
- The organization mistakes acceptance for approval, cooperation for trust, and technical credibility for social credibility.
- The organization does many of the following actions:
  - Fails to understand the local community's local rules of the game and is unable to establish social legitimacy
  - Delays the start of the stakeholder engagement process when there has been a significant change in its internal and external context
  - Fails to allocate sufficient time for relationship building
  - Undermines its credibility by failing to give reliable information or failing to deliver on promises made to the key stakeholders or the community
  - Underestimates the time and effort required to gain a social license to operate
  - Overestimates the quality of the relationship with the key stakeholders or the community

While the term "community" is used when talking about a social license to operate, it is really about the interests of the critical stakeholders, especially when they align with the community's norms and embrace its principles. Creating a relationship with stakeholders is more challenging when their interests are not in line with others in the community. Organizations need to have the ability to use techniques for capacity building at the local level. This is often accomplished with outside consulting help.

There are some cases where it is difficult to achieve a social license to operate. Like the practice of sustainability, there is no "one-size-fits-all" activity for either effort. It is always prudent to seek the counsel of people who understand the social structure and let them guide the organization as it performs its analysis of the external operating environment and applies this information within the context of its risk assessment process to rank order its opportunities and threats. A risk assessment process can be a valuable asset when an organization is involved in obtaining its social license to operate.

Finally, it is important to remember that the quality of the social license to operate is dynamic and responsive to changes in perceptions regarding the organization and its activities, decisions, and projects. It is also susceptible to outside influences. Therefore, the organization must diligently work hard to maintain its social license to operate over time once it is obtained.

There are many valuable publications dedicated to the social license to operate. Unfortunately, many of them are written for large corporations for use with operations that extract resources from the ground. Be careful that the information used by a specific organization is tailored to that organization and its stakeholders.

INTERNALIZING STAKEHOLDER INTERESTS IN AN ORGANIZATION

Organizations of all sizes and types can use risk management[8] to help embed the determination of the internal and external context into how they operate daily. The organization determines the internal and external context using the influences determined by the PESTLE and TECOP analysis methods. Context factors must be relevant to its purpose and affect its ability to achieve its responsibly set objectives. Furthermore, the context factors include conditions addressed by the three responsibilities within the organization's sustainability policy. Finally, the knowledge gained by understanding the context is considered when establishing and maintaining the organization's sustainability program.

UNDERSTANDING THE INTERESTS OF THE STAKEHOLDERS

The organization must determine the:[8]

- Stakeholders that are relevant to the achievement of its responsible objectives
- Relevant interests of the stakeholders
- Need for a social license to operate any portion of its operations, products, or services

The knowledge gained is considered when establishing and maintaining the organization's three-responsibility sustainability program. An effective stakeholder engagement process provides the lookout function, which monitors context so that objectives can remain responsible and the social license to operate can be maintained.

DETERMINING THE SCOPE OF THE SUSTAINABILITY PROGRAM

The organization determines the scope of the sustainability program by establishing its boundaries and applicability. When choosing the scope, the organization considers:[8]

- External and internal context
- Opportunities and threats associated with the uncertainty in the context
- Interests of the stakeholders
- Organizational functions and physical boundaries
- Need to obtain and maintain a social license to operate
- Authority and ability to exercise control or influence

The scope should include all activities, processes, products, and services within the organization's control or influence that can pose a significant risk to the organization as determined in the risk assessment process. Some organizations share the scope of their sustainability effort with internal stakeholders and many key external stakeholders.

SUSTAINABILITY FRAMEWORK

To continually improve its sustainability performance, an organization may seek to establish, implement, maintain, and continuously improve its sustainability framework as informed by the sustainability principles and process. The organization determines:[8]

- How it will satisfy its environmental, social, and economic responsibilities within the sustainability program
- How it will embed this framework into its operations—part of how the organization operates each day
- The influence of its stakeholders and their interests

Sustainability needs to be included within the scanning of the external operating environment and in the uncertainty analysis activity using stakeholder engagement (sometimes referred to as "communication and consultation") because:[8]

- The interests of internal and external stakeholders are essential to an organization's ability to meet its objectives.
- People will need to take (or not take) actions to manage sustainability effectively.
- People have some of the knowledge and information upon which effective sustainability management relies (i.e., obtaining the social license to operate).
- Some people might have a right to be informed or consulted about the activities and decisions of the organization.

Stakeholder engagement (i.e., communication and consultation with a two-way dialogue) is a critical supporting activity for all processes and decisions within the sustainability and risk management processes.

The organization should develop and implement a plan as to how it will communicate with the external stakeholders. This plan should include:[8]

- Engagement of the appropriate external stakeholders and ensuring effective exchange of information
- External reporting to comply with legal, regulatory, and government requirements
- Providing feedback and reporting on communication and consulting as part of the stakeholder engagement program

- Using engagement communication to build confidence within the organization
- Digitally communicating with stakeholders to make it convenient for them

These planning elements should include processes to consolidate risk management and sustainability information from all available sources.

Remember that engagement, communications, and consultation are processes, not outcomes. They usually take place with stakeholders. The beneficial effects of these processes are based on the transparent exchange of information and persuasion. Decisions are continuously improved by the engagement of stakeholders and keeping them digitally engaged. Stakeholder engagement is what separates sustainability from "business as usual."

CONCLUDING THOUGHTS

Stakeholder engagement is an essential and helpful component of every sustainability program. Failure to develop stakeholder engagement within the sustainability program can lead to the loss of the organization's social license to operate. In addition, an effective stakeholder engagement process can help reduce the effects of uncertainty (i.e., opportunities and threats) caused by the external operating environment.

It is worth reiterating the importance of a robust stakeholder engagement effort, especially when there is a good deal of change in the organization's context. Identifying and engaging stakeholders is how the organization manages its uncertainty to meet its strategic objectives. With this understanding, the organization can seek alignment with its stakeholders' interests while demonstrating how its activities, processes, decisions, products, and services can benefit these people and their organizations. An organization never deals with all the stakeholders simultaneously, but by following the VOC procedures with the stakeholders, the engagement will be separate and specific to particular interests. Some of the engagement will be driven by the stakeholders, while other engagement elements will be initiated and maintained by the organization itself.

Stakeholder engagement must be documented as part of the organization's knowledge management effort. This documentation should be maintained to help improve decision-making and help determine the degree to which sense-making needs to be actively scanning the external operating environment.

ESSENTIAL QUESTIONS FOR STAKEHOLDER ENGAGEMENT AND THE SOCIAL LICENSE TO OPERATE

1. How are the stakeholders identified within the scanning of the external operating environment?
2. Why is it essential to engage the external stakeholders to discuss their interests in the organization's sustainability efforts?

3. How does authentic engagement differ from other tools (e.g., surveys) used to understand the stakeholders' interests?

4. Why has the social license to operate concept not received greater use by organizations operating in a community setting?

REFERENCES

1. ISO (International Organization for Standardization), *Social Responsibility Guidance. ISO 26000*, Geneva, 2010.
2. P. M. Institute, *A Guide to the Project Management Body of Knowledge*, vol. 5th ed., Newtown Square: Project Management Institute, 2013.
3. M.A.M.D. Edmond, "Empowering the workforce to tackle the useful many processes," *Juran's Quality Handbook*, vol. 6th ed., New York: McGraw-Hill: In Juran, J.M., and DeFeo, J.A. Eds., 2010, p. 866–874.
4. G. Conlon, *Capitalizing on Voice of the Customer*, Stamford: Peppers & Rogers Group.
5. K. Dornerand D. Edelman, "What 'digital' really means," 2015. 2015. [Online]. Available: www.mckinsey.com/insights/high_tech_telecoms_internet/ what_digital_ really_means.
6. Geneva, *Quality Management Systems: Fundamentals and Vocabulary. ISO/DIS 9000*, Geneva: ISO (International Organization for Standardization), 2014.
7. I. Thompson and R. Boutilier, "The social license to operate. n.d. Website," Retrieved August 8, 2015, from http://sociallicense.com..
8. Geneva, *Risk Management: Principles and Guidelines. ISO 31000*, Geneva: ISO (International Organization for Standardization), 2009.

7 Organizational Governance

ABSTRACT

The system an organization uses to make and implement decisions to meet its objectives is organizational governance. Governance systems vary with the size and type of organization and the environmental, economic, political, cultural, and social context associated with its location. The systems can be formal or informal and are from the perspective of an organization. Governance should not be confused with corporate governance, which focuses on governing bodies (e.g., the board of directors). Instead, it is a system with a person or a set of trustees with the authority and responsibility for realizing the organization's objectives.

Organizational governance is present in every organization as one of its core functions. It provides the framework for decision-making at all levels of the organization. The leadership of the organization usually influences the decision-making. Governance enables the leadership to take responsibility for the risks associated with its activities and make decisions to embed sustainability into the organization.

ORGANIZATIONAL GOVERNANCE

Organizational governance provides the framework or rules, relationships, systems, and processes that enable the authority to be exercised and controlled within an organization.[1] This includes how the organization and its people are held accountable. If the governance is effective, it is more likely that the organization will be able to meet its objectives in an uncertain world.

Governance enables the leaders to direct and control the uncertainty associated with the organization's opportunities and threats.[1] Effective uncertainty management is essential for the organization to understand its risks, modify them as possible, and maximize its chance of achieving its objectives. Risk is always focused on meeting the objectives, which is facilitated by managing the opportunities and threats that contribute to or detract from the objectives. Poor governance increases uncertainty and leads to problems with the organization's long-term stability. Governance, management of uncertainty, and risk are highly interdependent.

It is essential to separate the organizational perspective from the focus of "corporate governance." Corporate governance is focused on the fiduciary responsibility of the board of directors.[2]

GOVERNANCE PRINCIPLES

There is no "one-size-fits-all" model for governance systems. However, some common elements that most leaders believe constitute "good practice" in governance.[3]

DOI: 10.1201/9781003255116-7

Governance principles have been developed by the Organisation for Economic Co-operation and Development (OECD)[4] and have been undergoing evolutionary changes to accommodate sustainability programs. The key to success is to have sustainability embedded in the decision-making process at all levels of the organization. This process has the following structural elements: commitment, governance policy, leadership responsibility for governance, and continual improvement.[3] These governance principles overlap with the sustainability principles in the following areas:[5]

- Accountability
- Transparency
- Ethical behavior
- Respect for stakeholder interests
- Respect for legal and other requirements
- Respect for international norms of behavior
- Respect for human rights

The purpose of combining these principles and embedding them in the decision-making process at all levels of an organization is to:[3]

- Enhance organizational effectiveness and performance
- Understand and manage uncertainties to minimize the threats and maximize the opportunities
- Enhance the competitiveness of the organization in the local community
- Strengthen the confidence of other organizations when a relationship is formed
- Enhance the public reputation of an organization through enhanced transparency and accountability
- Allow organizations to demonstrate how they are monitoring their ethical obligations
- Provide a mechanism for benchmarking the degree of organizational accountability with other organizations in the community
- Assist in the prevention and detection of fraudulent, dishonest, or unethical behavior

DEVELOPING THE GOVERNANCE SYSTEM

Organizations must innovate and adapt their governance practices to address the stakeholders' interests through sustainability and sustainable development to remain competitive in an uncertain world. These organizations must also grasp new opportunities to offset their threats. Management of the uncertainty effects must be a key to any governance program.[4]

The structural elements of governance include the following: commitment, governance policy, leadership responsibility for governance, and continual improvement. These elements help implement the organization's commitment to effective governance and address the common elements that constitute good governance practices over the long term. In addition, organizations typically use a formal or informal code of conduct to convey the governance policy.[6]

Some additional operational elements can improve direction for those with the responsibility for governance within an organization. The elements can include:

- Means to identify key governance issues depending on the context of the organization
- Operating procedures for maintaining governance as part of the daily operations of the organization
- Process for dealing with governance breaches and complaints
- Documentation of the governance program
- Internal reporting with links to the organization's strategic development, management of the uncertainty, and organizational objectives

To maintain the governance system over time, the organization must maintain awareness-building activity. The governance must maintain visibility and enable dialogue and communication within the organization. Governance must be carefully monitored to determine its effectiveness while continually committing the resources necessary to improve. As an organization grows, it needs to review the elements of governance regularly. Partnerships and other relationships, formal and informal, are informed of the results of any reviews that take place.[6]

Organizational governance is an essential factor in enabling an organization to embed sustainability. In addition, organizational resilience is an outcome of organizational governance.[7]

ORGANIZATION'S CODE OF CONDUCT

A code of conduct is a document that can help shape the sustainability culture of an organization. This document sets the standards of behavior expected of all members or employees in an organization.[8] It also helps them deal with ethical dilemmas they experience while serving in the organization. The code of conduct is used by the organization to:[6]

- Effectively deal with compliance with relevant laws and other rules
- Help management be more effective in monitoring the culture of the firm and its interactions with a wide range of different stakeholders
- Maintain the integrity and reputation of the organization and its members or employees

By establishing a culture of compliance, the organization can effectively limit its liability both in penalties and with its reputation.

A code of conduct should clearly state the organization's commitment to comply with laws and regulations. It must promote an internal culture of fair and ethical behavior and report matters that could be detrimental to the organization's reputation. This commitment can be quite informal in the case of small organizations. Nevertheless, a code of conduct is important in an organization's risk management program.[6]

The code of conduct applies to all the internal stakeholders. External stakeholders are often asked to comply with the code as a condition of their engagement with the organization, mainly when matters could impact the organization's compliance status or reputation. Many of the operational elements in the governance program should also be applied to the use of the code of conduct. Typically, the code of conduct covers environmental, social, and economic responsibilities to maintain a clear linkage with the sustainability program.[9]

In the true meaning of embedded sustainability, the governance, code of conduct, and three sustainability responsibilities (environment, social, and economic) are linked both within the organization and with the engagement with stakeholders. Sustainability is informed by the risk management process, the management of the uncertainty (i.e., the opportunities and threats), and several principles and processes. People and processes are vital to the long-term viability of an organization. Without all these elements present and aligned with each other, it would be difficult to fully embed sustainability, governance, and the code of conduct into a uniform decision-making framework used at every level of the organization.

Leadership

Effective leadership is critical to the governance system's success and the organization's ability to meet its strategic objectives. Leaders need to address the following actions:[10]

- Develop the mission, vision, values, and ethics while acting as a role model
- Define, monitor, review, and drive the continual improvement of the organization's processes and operations
- Engage directly with all stakeholders
- Reinforce a culture of operational excellence with the organization's members or employees
- Ensure that the organization is resilient to manage change effectively
- Some of the personal traits of leaders are as follows:[10]
- Ability to communicate a clear direction for the organization with a defined strategic focus
- Understand the management of uncertainty to enable the achievement of the strategic objectives of the organization
- Ability to be flexible in adapting and realigning the direction of the organization considering changes in the external operating environment
- Recognize that sustainable success relies on continual improvement, innovation, and learning
- Ability to use knowledge management and sense-making to improve the reliability of decisions at all levels of the organization
- Inspire while creating a culture of engagement, ownership, transparency, and accountability

- Be the role model for integrity, sustainability, environmental stewardship, social well-being, and shared value—both internally and in the community—in a manner that develops, enhances, and protects the organization's reputation

Leaders are responsible for creating the mandate and commitment to sustainability.

MANDATE AND COMMITMENT

Leaders of organizations need to demonstrate their commitment to governance with the following actions:[11]

- Ensuring that the risk management and sustainability policies are established following the organization's objectives, context, and strategic direction
- Ensuring that all the structural elements in this book are addressed with processes in all the operations of the organization
- Taking accountability for the effectiveness of the processes used and the efficiency of the operations associated with the organization's products and services
- Ensuring that risk management and sustainability are embedded in the process approach to the operations
- Communicating the importance of risk management and the responsibilities associated with sustainability to all stakeholders within the engagement process
- Ensuring that risk management and sustainability effectively contribute to enabling the organization to meet its objectives

The mandate aims to ensure that the organization clearly understands the benefits of risk management and sustainability and will embrace the change involved by embedding these practices into how the organization operates each day. In addition, the leader must be deeply involved in these programs to establish credibility with the external stakeholders.

Each organization should consider the following questions when establishing its mandate and commitment to risk management and sustainability:[12]

- How is sustainability embedded in the organization's objectives, and what goals and action plans are in place to help ensure success?
- Is the leadership clear about the significant opportunities and threats associated with the internal and external context and willing to engage with all stakeholders to ensure the organization can effectively manage the effects of uncertainty?
- Does the leader need to make changes to the prevailing risk attitude[13] that will facilitate embedding sustainability in the organization's activities, processes, and decision-making?

- Does the form of the policy that provides for embedding risk management and sustainability support the other policies that direct how the organization is operated?

Risk management and sustainability must be viewed as central to the "process focus" that describes the organization's operations.

Leaders must reinforce the organization's commitment and mandate through the following actions:[1]

- Make sure the risk management and sustainability objectives are carefully linked to the objectives derived from the mission statement
- Make it clear that risk management and sustainability are about effectively delivering the organization's strategic objectives with a uniform program of top-down objectives and a feedback loop of employee goals and action plans
- Ensure the risk management and sustainability activities required by the mandate are fully integrated into the governance and the organization's processes at the strategic, tactical, and operational levels
- Commit to making the necessary resources available to assist those accountable and responsible for managing risk and sustainability
- Require regular monitoring and measurement of the risk assessment and sustainability processes to ensure that they remain appropriate and effective
- Monitor to ensure that the organization has a current and comprehensive understanding of its opportunities and threats and that they are within the determined risk criteria
- Initiate corrective actions when threats are deemed to be unacceptable when they are not able to be offset by the opportunities
- Lead by example
- Review the commitment to the mandate as time, events, decisions, and the external operating environment conditions create change

Meeting the organization's objectives is essential to the organization. In addition, operating in an uncertain world makes it more critical that the leadership is committed to risk management and sustainability and that these practices are explicitly recognized in everything the organization does.

SUSTAINABILITY POLICY

Once the mandate is in place, leaders must make the mandate explicit through the issuance of risk management and sustainability policy. This policy aims to provide an effective means to express the organization's intentions and requirements clearly. The effectiveness rests on defining the organization's motivation for managing risk and sustainability and explicitly establishing what is required and by whom. Leaders must set the proper tone by emphasizing the necessity of using risk management and sustainability to help manage the opportunities and threats that define the organization's uncertainty. Short policies tend to be more effective. The policy should be succinct, as well as short in length.

Effective policy communication also requires confirmation that the information has been understood by the stakeholders inside and external to the organization. Stakeholders must genuinely believe that the policy reflects the organization's intent to embrace risk management and sustainability as part of its operations. To secure appropriate buy-in, the leader will need to produce tangible objective evidence that the organization has adopted changes in the way it has operated before adopting the risk management and sustainability policy and that those changes can persist over time. The organization's reputation and the credibility of its risk management and sustainability management efforts are dependent on its success in meeting the commitments embodied in the policy.

A risk management and sustainability policy is established, reviewed, and maintained by the leaders of an organization. Generally, it must meet the following requirements:[11]

- Be appropriate to the purpose and context of the organization
- Provide a framework for setting and reviewing objectives affected by the risk management and sustainability contributions
- Include a commitment to satisfy all applicable legal and other requirements
- Include a commitment to effective processes, efficient operations, and efficacious strategy
- Include a commitment to continual improvement, innovation, and learning
- Be maintained as documented information
- Be communicated within the organization
- Be a key part of the engagement with the external stakeholders

All reviews of the risk management and sustainability processes begin with examining the sustainability policy. It is the keystone element of all activities, processes, decisions, products, and services associated with the organization.

ORGANIZATIONAL ROLES, RESPONSIBILITIES, AND AUTHORITIES

As the organizational governance requires, leaders must ensure that the responsibilities and authorities for relevant roles are assigned, communicated, and understood within the organization.[11] Leaders need to assign responsibility for the following:

- Ensuring the risk management and sustainability processes are established with a reference or benchmarked to "best practice" for the organization
- Ensuring that the processes are effectively delivering their intended outcomes
- Reporting on the performance of the risk management and sustainability processes, on opportunities for improvement, and the need for change or innovation

- Ensuring the promotion of risk management and sustainability throughout the organization
- Ensuring the integrity of the risk management and sustainability processes when changes are planned and implemented in the organization

It is crucial to have a process focused on managing risk and sustainability that goes beyond preparing a sustainability report. Instead, the focus must be on identifying processes and operations that must be controlled to maintain and improve the organization's risk management and sustainability performance.

ORGANIZATIONAL STRATEGY

At the organization level, it is essential to set a strategy to meet the strategic objectives typically in a five- to ten-year time frame. In addition, the effects of uncertainty (i.e., opportunities and threats) must be managed for the strategy to be executed most effectively. A strategic plan provides a set of processes executed within the organization to meet the overarching objectives in an uncertain world. The focus is on developing strategies that will drive the organization to achieve the mission-derived objectives.

Organizations exist to create benefits for their internal and external stakeholders. The mission and the strategic objectives define these benefits. Products and services are designed within the operations, representing the organization's tactics. An organizational strategy helps the governance be accountable. This strategy also helps the leaders make decisions that support its execution. By binding together, the governance and the leadership, the organization is prepared for implementation (Figure 7.1). Most strategic failures are common failures of implementation.[14]

FIGURE 7.1   Strategy links governance and leadership in an organization.[14]

A strength, weaknesses, opportunities, and threats (SWOT) analysis helps organizations use their strength to take advantage of the opportunities and overcome the threats. This analysis also assists organizations in minimizing the weaknesses that are subject to threats.[15]

Strategic planning must account for the effects of uncertainty and how they affect the future. A scenario is an internally consistent view of the future. Scenario planning is generating and analyzing a set of different futures.[14] The results from building scenarios are not an accurate prediction of the future; instead, they provide better thinking about the future. Scenarios are used in strategic planning to provide a context for decisions. As events and incidents unfold in the external operating environment, it is necessary to review whether plans fit the realities of the PESTLE analysis results. The planning team needs to revise its scenario analysis if they do not.[14]

Execution of a strategy is a process. It is not an action or a step; it depends on more people than the strategy formulation. An organization navigates its strategy by focusing on the core building blocks (Figure 7.2).[16] It is best to move in small steps while maintaining a balance between strategizing and learning modes of thinking. Learning adheres to the same principles as the process of evolution.[14] Given the effects of uncertainty on the objectives, there is little alternative to adaptation. Only through action can an organization and its stakeholders participate and gather the experience that both sparks and are informed by the learning process.[14]

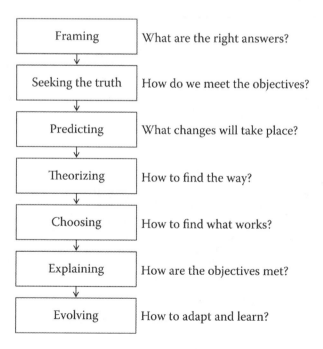

FIGURE 7.2 The building blocks of organizational strategy help organizations meet their objectives.[17]

A risk assessment is used to prioritize the opportunities and threats to ensure action that creates an efficacious strategy and strategic decisions that will enable the organization to meet its objectives. When there are frequent changes in the external operating environment, a risk assessment of the strategic options is required. It will facilitate the analysis of the stakeholder interests, customer requirements for products and services, competencies of the members or employees, and the PESTLE and TECOP analyses.[17] Strategies that are not executable are of no use. Essential questions for governance and leaders to address sustainability within the organization's strategy creation are as follows.

ESSENTIAL QUESTIONS FOR GOVERNANCE AND LEADERS TO CREATE AN EFFICACIOUS STRATEGY

1. How does an organization create a "system" by which it makes and implements decisions that will help it meet its strategic objectives?
2. How would a decision-making framework resemble the knowledge management process developed by Olivier Serrat (Asian Development Bank)?
3. Why is leadership a process rather than a personal trait or characteristic of a person who is a leader?
4. How are leadership and management alike and different from each other?

REFERENCES

1. ISO (International Organization for Standardization), *Risk Management: Guidance for the Implementation of ISO 31000. ISO/TR 31004*, Geneva: 2013.
2. O. Serrat, "A primer on corporate governance.," 2017. 10.1007/978-981-10-0983-9_52
3. Standards Australia International, "Good governance principles. AS 8000," 2003, [Online]. Available: https://www.saiglobal.com/PDFTemp/Previews/OSH/as/as8000 /8000/8000-2003(+A1).pdf.
4. OECD (Organisation for Economic Cooperation and Development), *OECD Principles of Corporate Governance*, Paris: OECD Publications Service, 2004.
5. ISO (International Organization for Standardization), *Social Responsibility Guidance. ISO 26000*, Geneva, 2010.
6. Standards Australia International,"Organizational code of conduct. AS 8002-2003. Sydney: Standards Australia International Ltd, 2003, [Online]. Available: https://www .saiglobal.com/PDFTemp/Previews/OSH/as/as8000/8000/8002-2003(+A1).pdf
7. BSI Group, "Guidance on organizational resilience. BS 65000," *BS (British Standards)*. London: British Standards Institute, 2014. [Online].
8. IFAC (International Federation of Accountants), *Defining and Developing an Effective Code of Conduct for Organizations*, Vols. Retrieved February 8, 2016, New York: IFAC, 2007.
9. Standards Australia International, "Corporate social responsibility. AS 8003," . Sydney: Standards Australia International Ltd, 2003.
10. EFQM (European Foundation for Quality Management)," *EFQM Excellence Model*, 2010.
11. ISO (International Organization for Standardization), *Environmental Management Systems: Requirements with Guidance for Use. ISO FDIS 14001*, Geneva, 2014.

12. ISO (International Organization for Standardization), *Risk Management: Guidance for the Implementation of ISO 31000. ISO/TR 31004*, Geneva, 2013.
13. SO (International Organization for Standardization), *Risk Management: Principles and Guidelines. ISO 31000*, Geneva, 2009.
14. O. Serrat, "From strategy to practice. Knowledge Solutions," Vols. Retrieved June 29, 2015, 2009.
15. J. a. M. G. Kiptoo, "Factors that influence effective strategic planning process in organizations," *Journal of Business Management*, vol. 16, no. 6, Vols. Retrieved February 8, 2016, p. 188–195, 2014.
16. C. D. A. a. M. A. Bradley, "Mastering the building blocks of strategy," Vols. Retrieved February 8, 2016, *McKinsey Quarterly*, October 2013.
17. P. Hopkin, *Fundamentals of Risk Management*, Philadelphia: Kogan Page, 2nd ed., 2012.

8 Uncertainty Assessment of Opportunities and Threats

ABSTRACT

Risk management is unique among other types of management because it specifically addresses the effect of uncertainty on objectives. Risk can only be assessed or successfully managed if the nature and source of that uncertainty are understood. Uncertainty in the organization's internal and external operating environment creates many opportunities and threats that must be prioritized for further consideration and management. Stakeholders may contribute additional opportunities and threats during the engagement process with the organization. Leaders encounter opportunities and threats in their decision-making. A risk assessment process can be used to plan for risk management and sustainability and determine the significant opportunities and threats identified in these activities.

Risk in an organization is associated with the prospects of meeting the strategic objectives. By managing the effects of uncertainty, there is a much more direct path between risk and objectives. Significant opportunities contribute positively to the objectives, while significant threats contribute negatively to the objectives. Risk management works best when opportunities offset threats, whereby significant opportunities are embraced, and significant threats can be controlled or otherwise mitigated. These processes operate in an organization and are not the same risk common to other applications and fields.

UNCERTAINTY ASSESSMENT PROCESS

The purpose of the uncertainty assessment process is to provide what is referred to as evidence-based information and analysis that assists in making informed decisions on how to manage the uncertainty associated with opportunities and threats.[1] Risk assessment methods have been slow to adapt to this need to handle positive and negative effects in the same analysis simultaneously. Therefore, we now refer to the uncertainty assessment process rather than the risk assessment method in Figure 8.1.

DEFINING THE CRITERIA

The uncertainty assessment process depends on the definition of criteria to provide terms of reference against which the significance of an opportunity or threat is evaluated.[2] Risk criteria provide the means to compare the opportunities and threats and

DOI: 10.1201/9781003255116-8

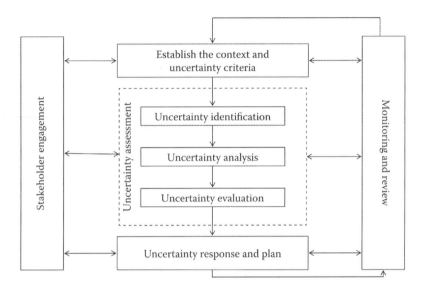

FIGURE 8.1   Structure for uncertainty analysis.[2]

determine their significance. Each organization creates its own procedure for determining significant opportunities and threats. The system needs to be accepted by the leaders. Once established, the risk criteria are applied throughout the organization whenever there are new opportunities and threats or when opportunities or threats are duly managed.[3] Risk criteria are often discussed within the stakeholder engagement process.

Risk criteria are generated in the following steps:[3]

1. List outcomes for each strategic objective: The desired outcomes for each organization's objectives should be identified. In the uncertainty analysis, these outcomes and how they are measured become the method of expressing the "consequence."

2. Select measures and scales for each outcome to characterize consequences: The organization needs to consider qualitative or quantitative measures that reflect the degree of success in achieving the objective, along with an appropriate scale on which to express each measure for each outcome. Consequences can be beneficial (objectives) or detrimental (threats). It is typical to have five levels for characterizing the consequences.

3. Decide how likelihood will be expressed: Likelihood can be measured in terms of probabilities (i.e., a value between 0 and 1) or frequencies (i.e., the number of occurrences over a unit of time, or by using descriptive scales). Likelihood scales relate to the likelihood of experiencing the consequences across a relevant time frame, the lifetime of a person, the expected life of an asset, the duration of a project, or government decisions affecting many generations. These ranges must be capable of expressing and distinguishing

consequences that are almost inevitable and highly likely from those that are expected to occur infrequently or are improbable.

4. Decide how consequences and likelihood will be combined to derive the level of uncertainty: The most straightforward way to combine consequences and likelihoods in qualitative risk analysis is to use a table to indicate the level of uncertainty the organization decides should correspond to each combination. Such a table is critical to the organization since it will determine how all opportunities and threats are evaluated. The table should be vetted within the stakeholder engagement and approved by the leaders.

5. Determine how the level of uncertainty will be expressed: Simple labels such as high, medium, and low can be used to represent the level of uncertainty, or a numerical scale can be used. To avoid confusion, it is essential that the terms (or numbers) used to describe the level of uncertainty are different from those used to describe levels of consequences or likelihood.

6. Decide on the rules for determining the significance of opportunities and threats: In most cases, the level of importance will be linked to an authority to accept and some indication of the need for embracing opportunities or avoiding threats.

ESTABLISH THE UNCERTAINTY CONTEXT

The term "context" is used differently. In an organization's management system, context refers to an understanding of the internal and external context of the organization. Understanding the context helps identify many of the opportunities and threats. For example, in uncertainty analysis, the purpose of the context is to reveal the sources of uncertainty that relate to the relevant objectives and the decision that the uncertainty analysis is seeking to make regarding the uncertainty response and plan.[4] The following questions should be asked when establishing the uncertainty analysis context:[5]

- What is the risk assessment and sustainability policy?
- What are the significant consequences and outcomes expected?
- What are the financial implications to the organization?
- What are the significant factors in the organization's internal and external operating environment?
- What are the related opportunities and threats?
- Who are the stakeholders associated with the opportunity and threats?
- What problems were identified in previous uncertainty analyses?
- What uncertainty criteria should be established for the analysis?
- What is the best way of structuring the opportunity and threat identification tasks?

Attention to these and other relevant factors should help ensure that the uncertainty analysis approach adopted is appropriate to the organization's needs and the risk associated with achieving the objectives.

STAKEHOLDER ENGAGEMENT

Communication and consultation for the uncertainty analysis occur within the stakeholder engagement process. In this respect, stakeholders should be internal and external to the organization. Managing uncertainty is important to people because:

- The interests of stakeholders are reflected in the establishment of responsible objectives for the organization
- People will need to take (or not take) particular actions for the opportunities and threats to be managed effectively
- People have considerable knowledge and information on which effective uncertainty analysis relies
- Some people might have a right to be informed or consulted

Stakeholder engagement is a key supporting activity for all parts of the risk assessment and sustainability program.

UNCERTAINTY IDENTIFICATION PROCESS

Uncertainty identification is a process used to compile a list of opportunities and threats that might contribute to or detract from achieving the organization's objectives. For each opportunity or threat, it is essential to know what, where, when, why, and how something could happen and the range of possible outcomes that affect the organization's objectives. In addition, it is important to link the factors and the specific sources of the opportunities and threats, along with their consequences (positive or negative). Events and leaders' decisions can also cause uncertainty and should be documented. The information on the opportunities and threats should be as complete as possible. It is usually compiled in a risk or uncertainty register. Failure to identify an opportunity or threat can result in an unexpected threat or a missed opportunity.

Uncertainty[4]

- Is a consequence of underlying sociological, psychological, and cultural factors associated with human behavior
- Is produced by natural processes that are characterized by inherent variability, for example, weather
- Arises from incomplete or inaccurate information, for example, due to missing, misinterpreted, unreliable, internally contradictory, or inaccessible data
- Changes over time, for example, due to competition, trends, new information, or changes in underlying factors
- Is produced by the perception of uncertainty, which may vary between parts of the organization and its stakeholders

An opportunity or threat is generally associated with the following components:[5]

- A source of the uncertainty
- An event, decision, or change in context
- A consequence, outcome, or impact on stakeholders of organization assets
- A cause (what and why), usually a string of direct and underlying reasons for the presence of an event, decision, or change in context
- Operational controls and levels of effectiveness
- The timing of the uncertainty and its likelihood

It is important to identify the sources of uncertainty and the interests of the internal and external stakeholders. An uncertainty analysis may concentrate on one or many possible areas of consequences (positive or negative) that may affect the organization. Information on the uncertainty must be sufficient to understand the likelihood and implications of the opportunity or threat. Existing information sources (e.g., stakeholder engagement and external context observations) must be accessed, and new data sources developed. Some uncertainties are challenging to observe.

Several questions need to be asked for each opportunity and threat:

- What is the source of each opportunity and threat?
- What might happen that could:
  - Increase or decrease the effective achievement of the organization's objectives?
  - Achieve the organization's objectives efficiently (financial, people, and time)?
  - Cause stakeholders to take action that may influence the achievement of the objective?
  - Produce additional benefits?
- What would the effect on objectives be?
- When, where, why, and how are these opportunities and threats likely to occur?
- Who might be involved or impacted?
- What controls presently exist to respond to the opportunities and threats?
- What could cause the control not to have the desired effect on the risk associated with the uncertainty?

As with all steps of the uncertainty analysis process, opportunity and threat identification should be iterative and repeated until a level of resolution and accuracy is obtained that is sufficient for the decisions concerned.

Naming opportunities and threats are crucial to effective uncertainty analysis and evaluation. Meaningful names should be given to the causes of opportunities, threats, causes, or initiating events to help talk about them. In addition, it is crucial to know how they might trigger an event and the consequences that might be associated with that event. Ideally, an opportunity or a threat should be identified in the following terms: (something happens) leading to (outcomes expressed in terms of its consequences concerning achieving the organization's objectives—positive and negative). The international risk management and uncertainty analysis do not recommend using a risk classification system.

Some organizations use a risk register to manage information on opportunities and threats. However, the document or database needs to be kept up to date. Risk and uncertainty change all the time. New opportunities and threats appear, and old ones disappear.

A risk matrix is also used to rank and display opportunities and threats by defining ranges for consequences and likelihood. Risk matrices should also be used with great care. The matrix doesn't need to be perfect and can crudely compare and rank opportunities and threats. However, it is important to note that a simple matrix can lead to the false impression of certainty about a consequence, its likelihood, or both.

The documented information from an opportunity and threat uncertainty determination should include the following:[3]

- Approaches and methods used
- Scope of the identification process
- Participants in the uncertainty identification and the information sources consulted
- Risk registers information on the opportunities and threats

Uncertainty identification establishes the information necessary for uncertainty analysis.

UNCERTAINTY ANALYSIS

Uncertainty analysis involves developing an understanding of the opportunities and threats. This provides input to uncertainty evaluation and decisions on whether threats must be treated, and opportunities must be embraced and developed. Uncertainty analysis also helps the organization make decisions regarding its options to deal with different types and levels of uncertainty and its effect on the level of risk. The results of the uncertainty analysis involve the preparation of a risk map in Figure 8.2.

Uncertainty analysis is a process of comprehending the nature of the opportunity and threat and determining the level of risk.[2]

Consequence is defined as[2] the outcome of the opportunities and threats, or an event or decision affecting the meeting of the organization's objectives. A means for qualitatively measuring consequences can be found in Figure 8.3.

An event (i.e., occurrence or change of the organization's context) can lead to a full range of consequences, each of which can be certain or uncertain and have positive or negative effects on objectives. Moreover, consequences can be expressed quantitatively or qualitatively.

A Likelihood is defined as[2] the chance of something happening. In uncertainty assessment, this term refers to the chance of something happening, whether defined, measured, or determined objectively or subjectively, quantitatively, or qualitatively,

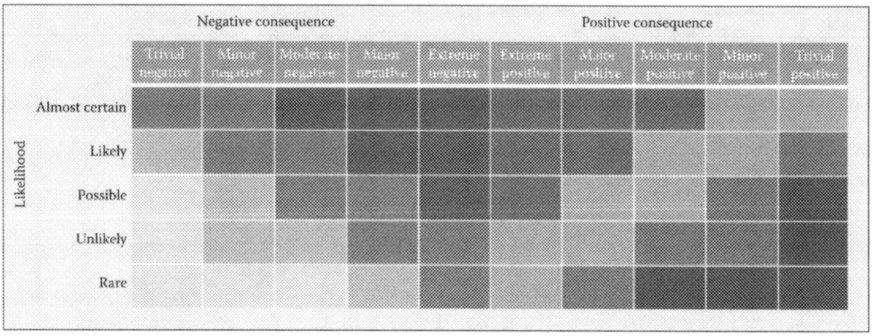

FIGURE 8.2 Risk matrix for the uncertainty analysis.[5]

Negative consequence	
Extremely negative	Most objectives cannot be achieved
Major negative	Some important objectives cannot be achieved
Moderate negative	Some objectives affected
Minor negative	Minor effects that are easily remedied
Trivial negative	Negligible negative impact upon objectives
Positive consequence	
Extremely positive	Most objectives should be achieved
Major positive	Some unimportant objectives cannot be achieved
Moderate positive	Some objectives affected in a positive sense
Minor positive	Minor improvement to chance of meeting objectives
Trivial positive	Negligible positive impact upon objectives

FIGURE 8.3 Determine the consequence of the uncertainty.[5]

and described using general terms or expressed mathematically. For example, a means for qualitatively measuring likelihood can be found in Figure 8.4.

Here are some key questions to ask during uncertainty analysis:

- What is in place that may prevent, detect, or lower the consequences or likelihood of the threats?
- What systems may enhance or increase the consequences or likelihood of opportunities or beneficial events?
- What are the consequences or range of consequences if they do occur?
- What is the likelihood or range of likelihoods of the opportunities or threats happening?
- What factors might increase or decrease the likelihood or the consequences?
- What additional factors may need to be considered beyond which the analysis does not hold?
- What are the limitations of the uncertainty analysis and the assumptions that are made?

Likelihood	
Almost certain	Consequence occurs often
Likely	Consequence occurs several times a year
Possible	Consequence occurs once every several years
Unlikely	Consequence occurs after decades
Rare	Heard of consequence occurring elsewhere

Qualitative analysis often used when the level of risk does not justify the time and resources needed to perform a numerical analysis. This is wellsuited for use in the analysis of opportunities and threats in the organization's context.

Semiquantitative analysis uses numerical ratings or more detailed descriptions for likelihood and consequence.

Quantitative analysis is used when the likelihood of occurrence and the consequence can be reliably quantified.

FIGURE 8.4   Determine the likelihood of the uncertainty.[5]

- How confident are you in your judgment or research, specifically about high consequences and low-likelihood opportunities and threats?
- What drives variability, volatility, or uncertainty?
- Is the logic behind the uncertainty analysis method sound?
- For quantitative analysis, what statistical methods may be used to understand the effect of uncertainty and variability?

Where there is a high level of uncertainty following the analysis, it may be appropriate to flag this and review the work at some future date and considering experience.

Qualitative uncertainty analysis is often used when the level of uncertainty does not justify the time and resources needed to perform a numerical analysis, where the numerical data is inadequate for quantitative analysis, or to perform an initial screening of uncertainty before the detailed analysis.[5] The value of qualitative uncertainty analysis is maximized when the uncertainty is determined and shared across many people with different backgrounds and interests. Collecting many ideas within the organization and throughout stakeholder engagement often improves the usefulness of the uncertainty analysis.

Semi quantitative uncertainty analysis uses numerical rankings (e.g., high, medium, and low) or more detailed descriptions of likelihood and consequences. This analysis can move to quantitative analysis when the possibility of occurrence and the consequences can be quantified.

UNCERTAINTY EVALUATION

The purpose of uncertainty evaluation is to assist in making decisions on the uncertainty response based on the outcomes of the uncertainty analysis. This enables an organization to determine which opportunities and threats must be responded to.

Uncertainty evaluation compares the level of uncertainty found during the analysis process with the uncertainty criteria established when the context was considered.

Decisions for responding to uncertainty should take account of the broader context of the uncertainty and consider the tolerance of opportunities and threats borne by stakeholders. Decisions should be made with full consideration of legal, regulatory, and other requirements. In some cases, the uncertainty evaluation can lead to conducting a more detailed analysis. The uncertainty analysis may also lead to a decision to respond to the opportunity or threat by only maintaining existing operating controls. Uncertainty analysis is influenced by the organization's risk attitude and criteria.[2] Based on this analysis, consider the need to avoid threats and embrace opportunities.

The generic strategy for uncertainty evaluation consists of four steps:[6]

1 Eliminate uncertainty: Seek to remove threats or offset the threats with opportunities, remembering that opportunities have risks associated with them should they not be able to be exploited successfully.
2 Allocate ownership: Seek to either transfer uncertainty to a third party to share in the effect of a loss for a fee or transfer ownership of an opportunity to a third party that can maximize the benefit with payment to the organization.
3 Modify exposure: In managing threats, this is referred to as mitigation, an attempt to make the likelihood or consequences smaller. When an opportunity is successfully implemented, it increases its likelihood and consequences on the upside, thereby avoiding the need to mitigate a threat.
4 Include in the baseline: Opportunities and threats that are not significant due to the uncertainty analysis can be put on a watch list to ensure that nothing is changing to make them more significant uncertainties. It is essential to consider multiple opportunities and threats that alone would end up in the baseline but, when clustered, may need some other form of response.

These four uncertainty evaluation categories can be looked at from the perspective of threats:[6]

- Avoid: Seek to remove threats to lower or eliminate uncertainty
- Transfer: Allocate ownership to enable effective management of a threat, often using an insurance company for this purpose
- Mitigate: Reduce the likelihood or consequence of the threat below an acceptable threshold
- Accept: Recognize residual risks associated with uncertainty and devise ways to control or monitor them

Uncertainty may be tolerated if the consequence or likelihood of that uncertainty is consistent with the established uncertainty criteria and their threshold of what the organization would consider unacceptable exposure. The ability of the organization to absorb an incident will, to a large degree, depend on its size and financial conditions.

Opportunity uncertainty response considers how to act to improve the likelihood and impact of an opportunity. When speaking about opportunity response options, four major categories of controls can be used:[6]

- Exploit identified opportunities, removing uncertainty by seeking to make the opportunity succeed.
- Enhance means increasing its positive likelihood or consequence to maximize the benefit of the opportunity.
- Share opportunities by passing ownership to a third party best able to manage the opportunity and maximize the chance of it happening.
- Ignore opportunities included in the baseline, adopting a reactive approach without taking explicit actions.

UNCERTAINTY RESPONSE

Uncertainty response involves selecting one or more significant opportunities and threats and exploiting the opportunities while avoiding the threats. In some organizations, avoiding threats is referred to as "risk treatment." Most organizations are still focused only on threats. In other organizations, leaders do not always view opportunities as being able to offset threats. There is a risk associated with funding efforts to take advantage of an opportunity. What if the opportunity is not realized despite attempting to take advantage of it? Without the implementation of the opportunity, it would be difficult to monetize the benefit associated with it. The constraint to developing opportunities as an uncertainty response is solely a function of an organization's experience and comfort in creating opportunities.

Several criteria have been defined to assess the effectiveness of these uncertainty responses:[6]

- Appropriate: Correct response level based on the uncertainty size.
- Affordable: The cost-effectiveness of responses should be determined.
- Actionable: Need to identify an action timeline since some uncertainty requires immediate attention while others can wait.
- Agreed: The consensus and commitment of the stakeholders should be obtained before creating the response.
- Allocated and accepted: Each response should be owned and accepted to ensure a single point of responsibility and accountability for implementing the response.

These criteria work well with both threats and opportunities.

MONITORING AND REVIEW OF PROCESS

Monitoring and review are essential steps in the process of managing uncertainty. It is necessary to monitor and review the effectiveness and efficiency of controls or execution and the appropriateness of uncertainty responses selected. In this way, it is possible to determine if the organization's resources were put to the best use and the uncertainty was reduced to enable the risks to be determined concerning the organization meeting its objectives.

The organization's monitoring and review processes should include:[2]

- Ensuring that controls are effective and efficient in both design and operation
- Obtaining further information to improve uncertainty analysis
- Analyzing and learning lessons from events (including near misses), changes, trends, successes, and failures
- Detecting changes in the external and internal context, including changes to uncertainty criteria and the risk itself—this may require revision of uncertainty responses and priorities
- Identifying emerging risks that lead to more uncertainty

Progress in implementing the uncertainty response should be measured. The results can be incorporated into the organization's performance program and the risk management and sustainability programs' internal and external reporting activities.

UNCERTAINTY MANAGEMENT PLAN

An uncertainty management plan provides a high-level model for how an organization should embed risk management into all its activities. This plan does not detail specific functional areas and activities of the organization. The plan may contain:[3]

- A statement of the organization's uncertainty management policy
- A description of the external and internal context, arrangements for governance, and the unique operating environment within which the organization operates
- Details of the scope and objectives of the uncertainty management activities in the organization, including organizational criteria for assessing whether opportunities and threats are to be avoided or embraced
- Uncertainty management responsibilities in the organization
- A list of opportunities and threats identified and an analysis of the opportunities and threats—usually in the form of a risk register
- Summaries of the uncertainty responses for significant opportunities and threats

The uncertainty management plan should provide fully defined and fully accepted accountability for opportunities and threats, controls, execution, and uncertainty responses for significant opportunities and threats.

Risk management and uncertainty analysis are certainly complicated at some level. The evaluation of traditional risk assessment methods tends to focus on threats and thereby not include the opportunities. Every organization will use some level of what is presented here to make better decisions regarding its opportunities and threats that were determined when preparing the internal and external context.

Please consider this information as a general program to guide any organization to have a reproducible means for prioritizing the opportunities and threats so that the activity to respond to opportunities and threats is within the organization's capability to support. Essential questions regarding the conduct of an uncertainty evaluation are as follows.

ESSENTIAL QUESTIONS ON UNCERTAINTY ASSESSMENT OF OPPORTUNITIES AND THREATS

1. How do the effects of uncertainty make it difficult to reliably determine the risk associated with the organization meeting its objectives?
2. Why does risk assessment need to be focused on the effects of uncertainty when risk is defined as the "effects of uncertainty" on meeting an organization's strategic objectives?
3. Why is it important to include the elements of the uncertainty analysis in the organization's stakeholder engagement process?
4. Why is it essential to include both the opportunities and threats in the uncertainty matrix?
5. How does the uncertainty evaluation help create the content for the uncertainty management plan?

REFERENCES

1. ISO (International Organization for Standardization), *Risk Management: Risk Assessment Techniques. ISO 31010*, Geneva, 2009.
2. ISO (International Organization for Standardization), *Risk Management: Principles and Guidelines. ISO 31000*, Geneva, 2009.
3. AS/NZS (Standards of Australia/Standards of New Zealand), *Risk Management Guidelines, Companion to AS/NZS ISO 31000:2009. HB 436*, Wellington, 2013.
4. ISO (International Organization for Standardization), *Risk Management: Guidance for the Implementation of ISO 31000. ISO/TR 31004*, Geneva, 2013.
5. K. Knight, "Risk management: A journey not a destination," 2010, Nundah, 2010. [Online]. Available: https://fermlab.hse.ru/data/2010/12/16/1208283693/A%20Journey%20Not%20A%20Destination%20-%20HO.pdf.
6. D. Hillson, "Effective strategies for exploiting opportunities," Presented at Proceedings of the Project Management Institute Annual Seminar and Symposium,, Vols. Retrieved February 25, 2016, 2001.

9 Sustainable Processes and Operations

ABSTRACT

Every organization focuses on improving the processes that constitute its operations to help meet its mission and strategic objectives, whether these processes are followed implicitly or explicitly. Effective processes help the organization manage the effects of uncertainty in its internal and external operating environment. To become sustainable over time, an organization must successfully identify and act on the opportunities contributing significantly to their strategic objectives infused with sustainability targets. The most effective way for an organization to optimize its opportunities is first to assess all elements of uncertainty with a focus on the organization's environmental, social, and economic responsibilities. These responsibilities form the foundation for meeting the intent of the sustainability policy. Second, the organization must affirmatively address the significant opportunities through the operational processes designed to support the strategic and operational objectives. Only in this way can the organization become more sustainable long term.

Efficient operations and processes are necessary to ensure the organization can achieve its mission and strategic objectives. Efficient processes, effective operations, and an efficacious sustainability strategy can be created and embedded in the way work is accomplished in the organization. These processes and operations can then significantly contribute to the organization's sustained success.

PROCESS APPROACH

An organization can achieve consistent and predictable results when its operations are understood and managed as interrelated processes. This approach to managing operations aligns all the processes to meet objectives. It is called a "process approach." A process is a set of interacting activities that use inputs to deliver a product or service. This process approach helps the organization:[1]

- Understand the process and be able to meet customer and stakeholder product and service requirements consistently
- Understand how the processes will add value
- Achieve effective processes and efficient operations
- Improve its processes based on the evaluation of data and information collected in the monitoring and measurement activities

The process approach is a management strategy for its product and service operations.[2] Using this strategy, the organization can develop, implement, and control its operational processes.[3] Here are some ways to establish a process approach for an organization's operations:[3]

- Determine and document product and service requirements.
- Establish a process to design and develop products and services.
- Monitor and control external processes, products, and services.
- Manage and control production and service provision activities.
- Implement arrangements to control product and service release.
- Control nonconforming outputs and document that all actions are taken.

While this list states what needs to be addressed, the organization can tailor the procedures to determine how they will be addressed.

WORKING WITH PROCESSES

Several widely used methods help organizations understand processes so the process approach can be put to good use. In the smallest organizations where operations are still implicit, these methods can be used qualitatively to understand processes. Other organizations can use these methods to create a straightforward process approach when meeting the objectives in the face of uncertainty becomes more urgent. Four such methods are described next.

TURTLE DIAGRAM

A turtle diagram provides a valuable means for understanding a single process (Figure 9.1). This diagram tracks the inputs and outputs of a single process. It also includes some of the critical dependencies that the process relies on:

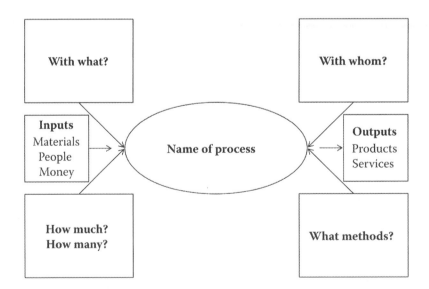

FIGURE 9.1   Turtle diagram for describing a single process.

- What are the required technology, machines, maintenance, materials, infrastructure, and work environment?
- How will it be measured, and how often?
- Who are the people and their requisite skills?
- What are the methods and procedures?

Many organizations prepare simple turtle diagrams to use for work instructions, and there are many versions available for inspiration by conducting an internet search. The example represented here is a generic concept.

PROCESS MAPPING

The hierarchical process mapping tool uses multiple layers to depict an entire system of processes responsible for a product or service (Figure 9.2).[4] It is easy to follow the flow since the high-level process has a single digit and a description of the work step. The second level has two digits, and the third level has three digits, and so forth. The supporting processes can be created using a turtle diagram and linked to the work steps. Hierarchical process maps are very useful for organizations to improve their processes.[5]

The recent use of process mapping, combined with Life Cycle Thinking, is used to identify waste in processes for organizations driving toward targets such as zero waste, net zero energy, or circularity. This process is easily adaptable and will assist in finding waste and hazardous substances in product processes.

SIPOC DIAGRAM

A supplier's inputs, process, outputs, and customers (SIPOC) diagram enable the organization to show how the suppliers and customers contribute to the entire value

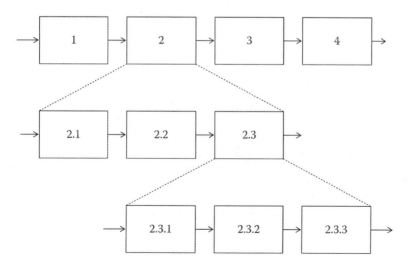

FIGURE 9.2 Multilevel hierarchical process map structure.[4]

FIGURE 9.3   SIPOC diagram—generic concept.

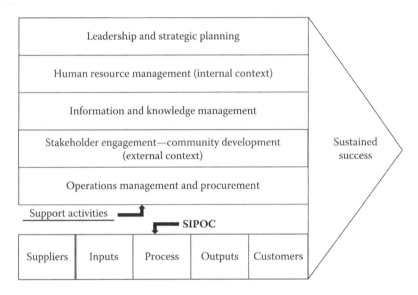

FIGURE 9.4   Value chain diagram—significant change.[6]

chain for a process (Figure 9.3). In today's global economy, every organization uses materials from many locations worldwide. Sustainability demands that every organization pay attention to their suppliers' environmental, social, and economic responsibilities. Many organizations are creating "supplier codes of conduct" for suppliers to ensure that all the interests are aligned. These same organizations are also receiving supplier codes of conduct from their customers. There is an expectation that these codes of conduct will be followed, and that each organization will use its sphere of influence to give meaning to sustainability throughout the entire value chain.

VALUE CHAIN DIAGRAM

The concept of the value chain[6] conveys the message that processes, and activities are universal to all organizations, whether they are a business or have some other purpose. An organizational value chain diagram depicts the organizational processes at the bottom of the diagram and the functional areas that can drive the performance of the processes at the top (Figure 9.4). It takes the combination of interactions between processes and "people processes" to create value and continually improve the processes so that the organization can meet its objectives.

MANAGING THE PROCESSES

- A plan–do–check–act (PDCA) method is used to manage processes and the organization's operations (Figure 9.5). The PDCA method can be described as follows:
- Plan: Establish the system of processes, along with the resources needed, to meet the mission, strategic objectives, and operating goals of the organization as it seeks to meet the customer requirements. Planning must also be sensitive to the stakeholders' interests and the opportunities and threats that were found to be significant in the uncertainty analysis.
- Do: Implement the system of processes as planned.
- Check: Monitor and measure the processes, along with the products and services, against the organization's policies, objectives, and requirements. The results of these checks need to be reported consistently with the organization's policies.
- Act: Take actions to improve process performance, as necessary.

All management systems can use this PDCA model to improve the processes continually.

According to the principles of risk management,[7] every organization needs to proactively manage its processes to have more effective and efficient operations. Organizations can use the process approach to establish formality in the more complicated processes, making it easier to determine interdependencies between processes, look for constraints, and see where resources can be shared. Processes need

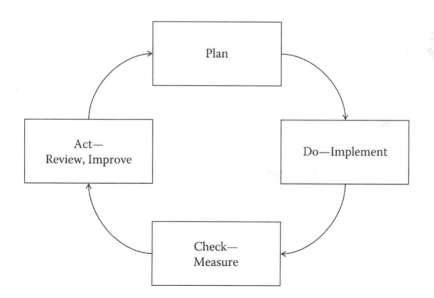

FIGURE 9.5   PDCA process—widely used generic concept.

to be reviewed regularly. When there are problems, the organization must take suitable actions to improve the processes and make them more effective.

As the organization becomes more complex, the processes need to be managed as a system whereby the networks of processes are created and understood in terms of their sequences and interactions. Consistent operation of such a system is referred to as a "systems approach" to process management.[8] Organizations can develop simple systems for managing the processes; however, these systems must become more rigorous when there are more people and greater demands on the process. This helps people understand what to do and what is an acceptable outcome.

PROCESS APPROACH AND CONTROL

Growth in organizations creates the necessity to accurately determine and plan the processes by which the organization will define the functions necessary for providing the products and services that will meet the needs and expectations of customers, organization members, employees, and other stakeholders on an ongoing basis. Planning these processes in light of the organization's strategy, mission, and objectives is essential. This planning enables the control of the processes by which the organization secures the provision of resources, facilitates the delivery of a product or service, monitors the activities, measures the performance of the process, and reviews the operating activities for ways to improve them continually.

Process planning and control within an organization help an organization in:

- Watching for changes in the external operating environment
- Forecasting the need for the outputs over the short and medium term
- Determining the interests of the stakeholders
- Understanding the objectives of the processes and how they get translated into the overarching strategic objectives
- Paying attention to the regulatory and legal requirements, as well as the three responsibilities of sustainability
- Managing the uncertainty by managing the opportunities and threats
- Controlling process inputs and outputs, including the use and loss (waste) of resources
- Paying attention to all interactions between processes
- Managing resource productivity and information and knowledge
- Monitoring activities and procedures, along with the people that are competent to perform the activities
- Maintaining records and other documented information required for transparency and accountability
- Conducting proper monitoring, measurement, and analysis of the process outcomes
- Taking corrective action when something does not work as planned
- Continually seeking improvement, innovation, and learning that benefit the outcomes of using the processes

Process planning must be attentive to the determined needs of the organization for developing or acquiring new technologies, developing new products or services, or both. This planning is how organizations add value that enables them to meet their strategic objectives.

Someone must be responsible for a process. Members or employees of an organization must be provided with defined responsibilities and authorities to complete the work. A "process owner" could be a person or a team of people, depending on the nature of the process and the culture and principles of the organization.

Processes need to be changed from time to time. These changes must be carefully planned to ensure they meet the requirements expected of the previous process while determining if any new or old conditions will no longer exist after the change. Just as changes in the external operating environment create new opportunities and threats, internal changes also introduce new opportunities and threats. Therefore, some internal changes may influence the external context as well. The information gathered by this process of managing change would be kept in the uncertainty (risk) register.

Many organizations restrict their planning activities to operations in a "normal" operating state, but what about emergencies and other uncertain operations or risk events? In an uncertain world, organizations are paying more attention to establishing procedures for determining how to respond to an internal incident, such as a fire, explosion, or accident. Organizations need to pay attention to how to:

- Respond to emergencies and accidents not covered by the characterization of the external context
- Take action to reduce the consequences of many different kinds of emergencies, appropriate to the magnitude of the incident and the potential for impact on the external stakeholders and the community at large
- Periodically test the procedures developed for managing these potential situations
- Periodically review and revise the procedures, where necessary, or do this after the occurrence of accidents, emergencies, or tests of the emergency planning system

It is important to remember that the operations are examined internally, and their associated opportunities and threats should be tracked in the uncertainty analysis.

RESOURCE MANAGEMENT

An organization must consider the internal and external resources required to achieve its strategic objectives in an uncertain world. These resources include facilities, equipment, materials, energy, water, knowledge, finances, and people. Resources must be used efficiently so the operations can accomplish "more with less." An organization should identify and assess the uncertainty associated with the securing of certain materials and people when they are needed. This might include scarcity of available resources, as well as mass sickness affecting the members or employees in the organization. The processes need to be optimized in a way that would cover some level of contingency for these and other similar considerations. At the same time, the organization should be vigilant in its search for alternative resources and new technologies. This will lead to a more agile and resilient organization.

Financial resources are reviewed periodically with the organization's financial advisors. This is part of maintaining sufficient financial resources to support activities required to meet the organization's strategic objectives. Improving the effectiveness of the processes and the efficiency of the operations can improve the financial situation by:

- Reducing process, product, and service failures and eliminating materials, waste, and lost employee time
- Eliminating the costs of compensation associated with any guarantees or warranties, legal costs (including mediation and litigation costs), and the costs of lost customers, markets, and reputation

Human resources are significant to the long-term success of an organization. Leaders need to create a shared vision and values within the internal context, enabling people to become fully involved in helping the organization meet its objectives. It is also critical to the organization's success to provide people with a suitable work environment that encourages personal growth, learning, knowledge transfer, and teamwork. Every organization needs to have a means for managing people through a planned, transparent, ethical, and socially responsible approach. It is important that the members or employees of the organization understand the importance of their contributions and meeting organizational objectives.

Infrastructure must be managed with effective processes so that its operation helps the organizations operate efficiently. Such an effort includes:

- Maintaining the dependability of the infrastructure needed by the organization
- Ensuring its safety and security
- Providing infrastructure elements needed for products and service provision
- Vigilantly pursuing its efficiency, cost, capacity, and effects on the work environment
- Lowering the impact of the infrastructure and assets on both the work environment and the natural environment
- Identifying and assessing the opportunities and threats associated with the infrastructure to make sure that these uncertainties are included in the scan of the internal and external operating environment and in the creation of potential contingency plans for dealing with events that may disrupt the infrastructure

The work environment must be suitable to achieve and maintain the sustained success of the organization, as well as the competitiveness of its products and services. The work environment should encourage productivity, creativity, and a sense of well-being for the people working in or visiting the organization's premises. It must also meet applicable statutory and regulatory requirements and address all applicable local standards. A suitable work environment is derived from a combination of human and physical factors that include:

- Opportunities for greater involvement in creating processes that help develop the potential of people in the organization
- Safety compliance standards
- Ergonomics
- Dealing with stress and other psychological factors
- Location of the organization
- Proper facilities for members or employees
- Maximization of efficiency and minimization of wastes
- Adequate control of heat, humidity, light, and airflow
- Proper hygiene and cleanliness while diminishing noise, vibration, and pollution
- Access to nature, preferably physically but at least visually

Attention to the built environment is critical to an organization's sustainability and should not be handled as a separate "bolt-on" set of initiatives. In a sustainable organization, the built environment should be designed not only for cost efficiencies but also to provide physical and emotional health benefits and preserve and protect the environment.

Natural resources need to be protected to ensure the availability of materials in the future to meet the requirements of customers and stakeholders. The organization needs to identify the opportunities and threats associated with the availability and use of energy and natural resources in the short- and longer-term. Appropriate consideration should be given to the stewardship of its products and services and the development of designs and processes to address these opportunities and threats. It is important to minimize the environmental, social, and economic impacts over the entire life cycle of its products and services, from design, manufacturing, and service delivery to product distribution, use, and end of life.

EFFICACIOUS STRATEGY

The organization's strategy consists of how it seeks to meet its objectives in an uncertain world and plans to achieve these objectives. The strategy should be clearly stated in a "looking forward" document. The strategy must have the power to achieve the desired outcomes to be efficacious. If clear strategic options are created and vetted through stakeholder engagement and uncertainty analysis of each operational objective, goal, and action plan, then the result will be an efficacious strategy. The uncertainty analysis helps prioritize all the identified opportunities and threats within the operating system, the internal context, and the external operating working environment. Stakeholder engagement and the effective management of uncertainty ensure that the strategy will be efficacious and that strategic decisions will reliably deliver the desired outcomes. Uncertainty analysis and management are critical tasks for creating an efficacious strategy.[9] Uncertainty in the internal and external environment may produce a wide range of different opportunities and threats. The nature of the uncertainty can change almost without notice in a highly dynamic situation. Organizations must be vigilant in capturing all the factors from the PESTLE and TECOP analyses and use strengths, weaknesses, opportunities, and threats (SWOT) and other tools to determine and assess the consequences of opportunities and threats. Paying close attention to the uncertainty analysis and the development and vetting of the strategy provides the best information needed by the organization to improve its decision-making and ability to manage effectively.

In an uncertain world, changes in events and other circumstances can reduce the strategy's success. But if they are included in the uncertainty analysis, then changes in the strategy can be initiated when the sense-making activity detects changes in the context. Then, the organization can determine what controls need to be implemented to optimize the consequences of these changes.[9]

EFFECTIVE PROCESSES

Effective processes mean that the processes are the correct ones for delivering what customers and stakeholders require. While processes may be efficient, this does not mean that they are accurate or the most effective processes that the organization could employ.[9] Developing more effective processes is the way organizations can satisfy customers and stakeholders. The operational system, and its processes and activities, provide how the organization will translate the efficacious strategy into successfully attaining its objectives. The importance of the operating system in meeting strategic objectives makes it very clear that it is critical to have effective processes.[9] Correct processes are built into the operating system and should never be managed by independent sustainability initiatives.

The uncertainty analysis can help the organization determine the proper response to significant opportunities and threats. When implemented, these operational controls must also be evaluated for effectiveness. If the uncertainty management is not comprehensive, the processes will not be effective. The overall intention is to ensure that processes are effective and that operations are always efficient. Effective processes provide how operations are changed, and strategy is realized.

EFFICIENT OPERATIONS

All organizations need to have efficient operations. The best way to ensure the ability of the organization to meet its objectives is to improve the efficiency of the operations. Especially in difficult times, the operating system must continue to be executed as efficiently as possible. As mentioned, delivering more efficient operations can be ensured by developing processes and activities requiring fewer resources. This creates value for the organization. On the other hand, there is no point in operating efficiently if those operations are based on incorrect processes or separately operated sustainability initiatives.[9] Uncertainty management is also relevant to maintaining an uninterrupted, efficient operating system. This continuity is a critical component of any sustainability program.

Efficient processes have little to do with "going green." Efficiency and effectiveness are fundamental requirements of an efficacious strategy. Efficiency, effectiveness, and efficacy have existed much longer than "green teams." The focus on having an award-winning sustainability report or sustainability program will not be the most important contributing factor to enabling the organization to meet its objectives in an uncertain world. However, the fact that the organization has an efficacious strategy, effective processes, and efficient operations will be a great story to highlight as a key strategy in the organization's success at meeting its environmental, social, and economic responsibilities and thereby achieving its sustainability goals. Essential questions for creating sustainable processes and operations are as follows.

ESSENTIAL QUESTIONS FOR CREATING SUSTAINABLE PROCESSES AND OPERATIONS INCLUDING CIRCULARITY

1. Why is it important to manage an organization's operations using the ISO process approach?
2. How does an organization create effective processes throughout its operations?
3. How does an organization ensure that its operations are efficient?
4. Why does the sustainability strategy need to be efficacious?

REFERENCES

1. ISO (International Organization for Standardization), *Guidance on the Concept and Use of the Process Approach for Management Systems. ISO/TC 176/SC2/N 544R3*, Vols. Retrieved June 22, 2015, Geneva, 2008.
2. Praxiom, "ISO's process approach translated into plain English. Calgary, AB," Praxiom Research Group Ltd., 2015. [Online]. Available: http://www.praxiom.com/process -approach.htm.
3. Praxiom, "ISO FDIS 9001 2015 translated into plain English. Calgary, AB," 2015. [Online]. Available: http://www.praxiom.com/iso-9001.htm.
4. R. Pojasek, "Understanding processes with hierarchical process mapping," *Environmental Quality Management,* vol. 15, no. 2, p. 79–86, 2005.
5. R. Pojasek, "Putting the hierarchical process maps to work," *Environmental Quality Management*, vol. 15, no. 3, p. 73–80, 2006.
6. M. Weil and Kurt Ernst, "PORTER, Competitive advantage Creating and Sustaining Superior Performance," 1985: 82–84.
7. ISO (International Organization for Standardization), *Risk Management: Principles and Guidelines. ISO 31000*, Geneva, 2009.
8. R. Pojasek, *Making the Business Case for EHS*, Old Saybrook: Business and Legal Reports, 2005. [Online].
9. P. Hopkin, *Fundamentals of Risk Management*, London: Kogan Page Limited, vol. 2nd ed., 2012.

10 Organization's Support Operations

ABSTRACT

People are a critical resource in an organization. Therefore, leaders need to create and maintain a shared vision, shared values, and an internal context. This helps people in the organization become fully engaged in its operating environment and in meeting its strategic objectives, thus helping them understand the importance of their contributions and roles. Organizations must ensure that the internal context and work environment encourage personal growth, learning, knowledge transfer, and teamwork. The organization must perform the management of people in a planned, transparent, ethical, and socially responsible manner. People are involved as operators of the core processes and the supporting processes. However, many organizational support operations do not get the attention of the sustainability program. Without the range of different support processes, the core processes will not be able to function effectively.

SUPPORTING HUMAN RESOURCES

People are a significant resource of an organization, and their full involvement enhances their ability to create value for all stakeholders. Therefore, leaders should create and maintain a shared vision, shared values, and an internal environment in which people can become fully involved in achieving the organization's strategic and operational objectives.[1] People management should be performed through a planned, transparent, ethical, and socially responsible approach. In addition, the organization should ensure that the people understand the importance of their contribution and roles as a key part of any sustainability program.[1]

ENGAGEMENT OF PEOPLE

Competent and engaged people are essential to the organization's ability to meet its strategic objectives. Involved people at all levels of an organization can prove beneficial in the following ways:[2]

- Improved understanding of the organization's strategic objectives
- Increased motivation to help achieve the organization's objectives
- Increased involvement in improvement, innovation, and learning activities
- Enhanced personal development and creativity
- Higher levels of satisfaction
- Increased levels of trust and collaboration
- Increased attention to shared values and culture throughout the organization

DOI: 10.1201/9781003255116-10

To obtain these benefits, the organization must establish processes that engage people to:[1]

- Set the right goals and action plans to enable the attainment of organizational objectives
- Identify constraints to setting and achieving these goals
- Take ownership and responsibility for solving problems
- Assess personal performance against the goals and action plans
- Actively seek opportunities to enhance competence
- Promote teamwork and encourage collaboration among people
- Share information, knowledge, and experience within the organization

COMPETENCE OF PEOPLE

To ensure that the organization's members have the necessary competences, the organization should establish and maintain a personnel development plan and associated processes to implement the plan. The plan should assist in identifying, developing, and improving the competence of people through the following steps:[1]

- Identifying the professional and personal competences the organization could need in the short and long term
- Identifying the competencies currently available in the organization and the gaps with what is needed in the future
- Implementing actions to improve or acquire competences to close the gap
- Reviewing and evaluating the effectiveness of actions taken to ensure that the necessary competencies have been acquired
- Maintaining competences over time as needed

The organization needs to determine the knowledge necessary to operate its core processes that produce products and services. This involves its current knowledge and ability to acquire or access the required knowledge as the processes and operations are improved and maintained in a compliant condition. Some of this knowledge is derived from internal sources (e.g., learning from successes and failures), and some from external sources (e.g., standards, schooling, conferences, and engagement with customers and other stakeholders). Knowledge is a key element that supports decision-making at all levels of the organization.[1]

Next, the organization must determine the necessary competence needed for working within the management system and affecting organizational performance and effectiveness. It is important to ensure that people within the organization are competent based on their education, training, and experience. When necessary, actions should be taken so that people can acquire the necessary competence and that leaders can evaluate the effectiveness of these actions.

People working in an organization need to be aware of the policies, objectives, and goals that relate to their work and how they contribute to the effectiveness of the management system. This awareness helps people understand the benefits of using their knowledge and experience to help the organization meet its strategic objectives. Awareness also helps people understand the negative implications of not conforming to the management system.[1]

People learn about the progress toward meeting the strategic objectives through effective, two-way dialogue and communication. Therefore, leaders need to provide an appropriate emphasis on the process of engagement. Providing only mechanistic internal and external communication is not as effective or enduring.

INTERNAL VALUE CHAIN SUPPORT

The internal value chain has many functions that support the operations with processes associated with the activities of people. It is essential to have these people-focused support processes be as effective as those used in operations (i.e., which are directly associated with products and services). These support activities include:

- Leadership
- Strategic planning
- Employee engagement
- Stakeholder engagement
- Information and knowledge management
- Operations management

These support activities are found in many performance-focused programs. Each aforementioned category has criteria representing some level of "best practice." Organizations can self-assess against the criteria to compare their support processes against the best practices of organizations in each category.[3]

COMMUNICATION

The organization should determine the need for internal and external communication. It is helpful to view communication as the "messages" that inform the ongoing engagement process. Engagement is a form of dialogue not restricted to communication. Sustainability must be integral to this communication and not be conducted separately. A communication program should consider:[4]

- What it will communicate
- When it will communicate
- With whom it will communicate
- How it will communicate

Internally, the organization should adopt appropriate methods of communication to ensure that the sustainability message is heard and understood by all employees on an ongoing basis. Much of the communication provides information on the core and supporting operations. The communication should set out the organization's expectations of employees. External communication should give messages to the stakeholder engagement process. The entire range of stakeholders should receive all relevant external communications.

Communication should be designed to enhance the awareness of all the organization's stakeholders. This awareness should include:

- The sustainability policy and other policies of interest to each party
- Relevant strategic, operational, and tactical organizational objectives
- Their role and contribution to effective processes, efficient processes, and efficacious strategy
- The implications for not conforming to the system of management requirements

SUPPORTING INFORMATION RESOURCES

The organization should establish and maintain processes to gather reliable and valuable data and convert that data into the information needed for decision-making. Information support includes the processes required for storing, security, protecting, communicating, and distributing data and information to all engaged stakeholders.[1] An organization's information and communication systems need to be robust and accessible to people making decisions at all levels of the organization. The leaders need to do what is necessary to ensure the integrity, confidentiality, and availability of information related to the organization's performance, process improvements, and progress toward achieving sustainability.[1]

COMPLIANCE INFORMATION

Every organization must determine and provide the resources necessary for establishing, developing, implementing, evaluating, maintaining, and continually improving all environmental, social, and economic compliance information.[5] The information and how it is managed are related to the size, complexity, structure, and core and supporting processes. Leaders are expected to make the necessary compliance-related resources available and deploy them effectively to ensure compliance management meets the organization's strategic and operational objectives and achieves compliance.[5] These resources include financial and human resources and access to expert and skilled people who can consult with the organization based on their knowledge, professional development, and information technology. These experts may have legal, engineering, management systems, process improvement, experience credentials, and related information technology.

MONITORING AND MEASURING RESOURCES

The organization should identify the internal and external resources needed to achieve the organization's strategic and operational objectives in the short and long term. To ensure that these resources (e.g., equipment, facilities, materials, energy, knowledge, finances, and people) are used effectively and efficiently, the leaders need to have processes in place to provide, allocate, monitor, evaluate, optimize, maintain, and protect those resources.[1] To ensure the availability of the resources for future activities, the organization needs to identify and assess the threats of potential scarcity and continually monitor the current use of economic, material, and people resources to find opportunities for the optimization of their use. This should be done parallel with research for new resources, optimized processes, and technologies.[1] The organization needs to monitor and measure the availability and suitability of the identified resources, including outsourced resources, and act as necessary. The results of the monitoring should also be used as inputs to the process of maintaining the organization's efficacious strategy.

Organizations use monitoring and measuring to ensure that valid and reliable processes are used for their products and services. In some cases, these resources include measuring equipment[4] that is suitable for the activities involved and maintained to ensure fitness for its purpose.

In some processes, there is a requirement for measurement traceability, especially when it is considered an essential part of providing confidence in the validity of the measurements.[4] Measurement standards are used for this purpose. When no international or national measurement standard exists, the basis used for calibration or verification is retained as documented information. All measuring equipment and standards are protected from adjustments, damage, or deterioration that would invalidate the instrument calibration and subsequent measurement results.[4]

FINANCIAL AND RISK MANAGEMENT INFORMATION

Leaders should determine the organization's financial needs and establish the necessary financial resources for current and future operations.[1] The organization needs to establish and maintain processes for monitoring, controlling, and reporting the effective allocation and efficient usage of financial resources related to the organization's strategic objectives. Reporting the financial information can also provide the transparency required for determining ineffective or inefficient activities while receiving feedback on the economic justification for initiating suitable improvement actions. Financial reporting should be used in all management reviews and evaluations of the effectiveness of the sustainability efforts. Improving the effectiveness and efficiency of the organization's operations and supporting core operations can positively influence the financial results.

There are areas where supporting processes and the mechanical and information assets need to be carefully monitored by the organization's financial and risk management (e.g., insurance provision) functions.[6] A key element in sharing information involves life cycle costing analysis and the costing levels of service options. This information is required for benefit analysis that will be used to evaluate the entire set of operations—core operations and supporting operations.

Many insurance carriers seek information on specific equipment or practices that could lead to financial exposures. The concerns arise with many of the issues mentioned in this chapter. By managing these concerns, the processes should be more effective, and the amount of money spent on insurance should decrease over time.

KNOWLEDGE MANAGEMENT INFORMATION

An organization needs to identify the knowledge necessary to improve the effectiveness of the processes and the efficiency of the core and supporting operations. The knowledge management processes should address how the organization identifies, obtains, maintains, protects, uses, and evaluates the need for these knowledge resources. Leaders must assess how the organization's current knowledge base is identified and protected. Leaders should also consider how to obtain the knowledge required to meet the present and future needs of the organization from both internal and external resources.[1]

There are many ways to identify, maintain, and protect knowledge:[1]

- Learning from failures, near-miss incidents, and successes
- Capturing the knowledge and experience of the people in the organization
- Gathering knowledge from customers, stakeholders, suppliers, and partners
- Capturing undocumented knowledge (tacit and explicit) existing in the organization
- Capturing important information from documented data and records

This information is vital to their ability to offer conformant products and services. Knowledge is also required to improve decision-making at all levels in the organization.

DOCUMENTED INFORMATION

Documented information is created by the organization based on what is necessary to ensure the effectiveness of the management system. This will differ substantially with the size of the organization and its activities, processes, products, and service, as well as with the competence of the people within the organization.[4]

Creating and updating documented information needs to be performed consistently and reliably. There is a need for the identification and description (e.g., title, date, author, or reference number), a functional format (e.g., wording, software version, and graphics), and the periodic review and approval of the documentation for suitability and adequacy.[1]

Finally, documented information needs to be carefully controlled to be available where and when required and protected (e.g., from improper use, loss of confidentiality, or loss of integrity).[1] To properly control documented information, the organization needs to address its distribution, access, retrieval, and use. In addition, there must be a provision for proper storage and preservation of the documents. It is important to maintain both version control and legibility of the documents, along with plans for the retention or destruction of the documents when it is deemed they have served their useful or legal purpose. Some documents often come from outside the organization (e.g., standards, customer methods requirements, and service manuals). The degree to which these are managed needs to be determined and assessed by leadership and included as part of the control function.[1]

INFORMATION ON PHYSICAL ASSETS

The organization needs to provide the infrastructure necessary to support the operations by planning, providing, and managing its infrastructure effectively and efficiently to help meet its operational objectives. In addition, the infrastructure supports the organization's ability to provide products and services. Examples of infrastructure include the following:

- Buildings; storage areas; tank farms; roadways; electricity; water; compressed air; fire suppression; compressed gases; heating, ventilating, and air-conditioning; and filters
- Equipment, including surveillance systems, computers, microprocessors, and software
- Transportation, including loading docks, forklifts, containers, trucks, conveyors, and elevators
- Information and communication technology, including speaker systems and alarm systems

Infrastructure is maintained as part of maintaining control in the assets management system.

The process of managing the organization's infrastructure should provide consideration of the following:[1]

- Dependability of the infrastructure
- Safety and security
- Efficiency, cost, capacity, and work environment
- Impact of the infrastructure on the work environment

Organizations need to identify and assess the risks associated with the infrastructure and take appropriate action to manage uncertainty, including establishing adequate preparedness and response plans.[7]

Leaders should consider technology options to enhance the organization's performance in operations, supporting operations, marketing, benchmarking, customer and stakeholder interaction, supplier relations, and all outsourced processes.[1] The organization should establish effective processes for the assessment of:[1]

- The current levels of technology usage inside and outside of the organization, including emerging trends
- Economic costs and benefits
- The evaluation of opportunities and threats related to changes in technology

- The competitive environment
- Its speed and ability to react to customer and stakeholder interests

WORKING ENVIRONMENT

The organization must determine, provide, and maintain the work environment necessary to support the quality of the products and services. In addition, the work environment should encourage productivity, creativity, and well-being for the people working on or visiting the organization's premises. Leaders should seek to provide a suitable work environment to achieve and maintain the sustained success of the organization and to support the competitiveness of its products and services. Several human and physical factors need to be addressed:[1]

- Optimal social conditions determined by the social well-being responsibility of the sustainability program, including human rights, non-discrimination, proper discipline actions, safety rules, and guidance for the use of protective equipment
- Facilities for the people in the organization
- Physiological factors, such as stress reduction and fatigue recognition
- Ergonomic issues, such as repetitive motions
- Physical elements, such as heat, cold, humidity, light, airflow, personal hygiene, noise, vibration, and pollution
- Maximization of efficiency and minimization of waste

Environmental, occupational health, and safety requirements regulate many of these factors and situations. The factors in the workplace differ substantially depending on the products and services provided and the internal and external context of the organization.

AVAILABILITY OF NATURAL RESOURCES

The availability of natural resources is a critical concern in a sustainability program. The organization needs to consider the opportunities and threats related to the availability and use of energy, water, and other natural resources in the short and long term.[1]

An organization needs to consider the integration of environmental protection aspects into product and service design, as well as the development of its processes to address opportunities and threats. It is also essential to minimize environmental impacts over the entire life cycle of its products, services, and infrastructure, from design, manufacturing, or service delivery to product distribution, use, disposal, or other end-of-life options.[1] Essential questions for developing support processes and people are as follows.

ESSENTIAL QUESTIONS ON SUPPORT PROCESSES AND PEOPLE

1. How do people support the various processes responsible for producing the organization's products and services?
2. Why do the value chain support personnel need to pay attention to the engagement and competence of people responsible for processes that deliver products and services?
3. How does the value chain support personnel and help manage the supporting information resources?
4. How does the value stream support personnel and help manage the physical assets and the working environment?

REFERENCES

1. ISO (International Organization for Standardization), *Managing for the Sustained Success of an Organization: A Quality Management Approach. ISO 9004*, Geneva, 2009.
2. ISO (International Organization for Standardization), *Quality Management Systems: Fundamentals and Vocabulary, ISO/DIS 9000*, Geneva, 2014.
3. R. A. H. J. Pojasek, "Improving sustainability results with performance frameworks," *Environmental Quality Management*, vol. 20, no. 3, p. 81–96, 2011.
4. ISO (International Organization for Standardization), *Quality Management Systems: Requirements. ISO 9001*, Geneva, 2015.
5. ISO (International Organization for Standardization), *Compliance Management Systems: Guidance. ISO 19600*, Geneva, 2014.
6. ISO (International Organization for Standardization), *Asset Management: Management Systems: Requirements. ISO 55001*, Geneva, 2014.
7. ISO (International Organization for Standardization), *Risk Management: Principles and Guidelines. ISO 31000*, Geneva, 2009.

11 Organizational Sustainability

ABSTRACT

Successful implementation of organizational sustainability must be developed from within the organization by building on what is already in place. Organizational sustainability should create relationships between processes and operations, having regard for the stakeholders' interests. The organization incorporates sustainability within its management system to encourage resource productivity, decision-making, products, and services, as well as the interests of the stakeholders and the community. Organizations should seek an appropriate level of sustained success in line with the complexity of the decisions that must be made. In addition, an organization needs to understand its opportunities and threats and how they may influence its overarching strategic objectives. Ultimately sustainability should be embedded in the decision-making process by including sustainability in the organization's policies, culture, strategies, and operations.

ATTRIBUTES OF ORGANIZATIONAL SUSTAINABILITY

There are eight attributes that organizations can use to evaluate the extent to which sustainability is practical and whether it can be improved. First, organizations should aim at an appropriate level of performance of their risk management and organizational sustainability programs that align with the complexity of the decisions that need to be made.[1] The list of attributes represents organizational sustainability at a high level.

ETHICAL BEHAVIOR

An organization should behave ethically. This behavior stressed values of honesty, equity, and integrity. These values imply a concern for people, animals, plants, and the environment, as well as a commitment to address the impact of its activities and decisions in light of its stakeholder's interests.[2] This attribute can be tested by examining how the organization defines and communicates the standards of ethical behavior expected from its governance structure, personnel, suppliers, contractors, and owners and leaders. Usually, this information is found in the organization's "code of conduct," or it may be found in the customer's "supplier code of conduct," which is often made part of the purchase order for products and services.

RESPECT FOR THE RULE OF LAW

An organization must demonstrate its respect for the rule of law. It is generally implicit in the rule of law that laws and regulations are written, publicly

DOI: 10.1201/9781003255116-11

disclosed, and fairly enforced according to established procedures. In the context of sustainability, the organization should be aware of applicable laws and regulations and inform those within the organization of their obligation to observe and implement those measures.[2] This is tested by determining how the organization keeps itself informed of all legal and other obligations and ensures that its relationships and activities comply with the intended and applicable legal framework. The organization should also conduct a periodic evaluation of its compliance with applicable laws and regulations, including contractual adherence to codes of conduct.

RESPECT FOR STAKEHOLDER INTERESTS

An organization must respect, consider, and respond to the interests of its stakeholders. Besides the internal stakeholders, other individuals or groups may also have rights, claims, or specific interests that need to be considered. Sustainability also demands that the organization consider stakeholders' interests who may be affected by a decision or activity, even if they have no formal role in the stakeholder engagement efforts or are unaware of these interests.[2] This is tested by the ability of stakeholders to contact, engage with, and influence the organization. Therefore, there should also be documented information on stakeholders' role in conducting the organization's uncertainty analysis.

FULL ACCOUNTABILITY FOR OPPORTUNITIES AND THREATS

Uncertainty analysis includes comprehensive, fully defined, and accepted accountability for testing opportunities and threats for significance and taking the proper response. All members or employees of an organization must be fully aware of the opportunities and threats, the controls, and the engagement tasks for which they are accountable. The definition of risk management and sustainability roles, accountabilities, and responsibilities should be part of all the organization's work instructions and operating controls.[1] Many organizations include the performance of uncertainty assessment and sustainability accountabilities in performance reviews.

ENGAGEMENT WITH STAKEHOLDERS

Risk management and sustainability require continual, interactive communications with external and internal stakeholders, including comprehensive and frequent reporting of uncertainty assessments (i.e., opportunities and threats) and applying risk management and sustainability appropriately. Communication is seen as a two-way process so that properly informed decisions can be made about the level of uncertainty and the need for opportunity and threat response in line with properly established and comprehensive uncertainty criteria.[1] This can be tested by looking at the documented information on the stakeholder engagement process.

EMBEDDING SUSTAINABILITY IN THE ORGANIZATION'S GOVERNANCE

Risk management and sustainability are considered central to an organization's management system, so risks of meeting the organization's strategic objectives are improved by managing the effects of uncertainty (i.e., opportunities and threats). The governance structure and process are based on managing risk, uncertainty, and sustainability. Leaders regard effective management of opportunities and threats as essential for achieving the organization's strategic objectives.[1] in an uncertain world. This is indicated by leaders' language and the written materials in the organization using the term uncertainty in connection with risks to the strategic objectives.

INCLUDING SUSTAINABILITY IN DECISION-MAKING

All decision-making within the organization, regardless of its level of importance or significance, involves the explicit consideration of the effects of uncertainty (i.e., opportunities and threats) and the risk management of meeting its strategic objectives. This can be tested to determine if performance evaluation criteria include sustainability in the decision-making of individuals in the organization. There should be documented information to determine whether decision-making is expected to include explicit consideration of the organization's strategic objectives, the effects of uncertainty, organizational knowledge, risk management, and sustainability.

CONTINUAL IMPROVEMENT AND LEARNING

An emphasis is placed on continual improvement in risk management and sustainability through setting goals and action plans against which the organization's members assess whether the strategic and operational objectives are being met. The organization's performance can be documented and shared with the stakeholders[1] during the engagement process. Sustainability often has a regularly scheduled management review meeting as a component of its management system. The learning is expressed as the "act" component of the plan–do–check–act cycle.

SUSTAINED SUCCESS OF AN ORGANIZATION

Leaders should adopt a risk management and sustainability approach to achieve sustained success. The organization's management system should be based on sustainability principles; these principles describe concepts that are the foundation of effective risk management and sustainability programs.

The organization achieves sustained success by consistently addressing the interests of its stakeholders, in a balanced way, over the long term.[3] An organization's internal and external operating environment is ever-changing and uncertain. Therefore, to achieve sustained success, the organization's leaders should:[3]

- Maintain a long-term planning perspective
- Constantly monitor and regularly analyze the organization's operating environment
- Identify all stakeholders and assess their potential impacts on the organization's performance
- Engage stakeholders and keep them involved in the review of the organization's management of its significant opportunities and threats
- Establish mutually beneficial relationships with members of the organization's value chain
- Conduct uncertainty assessments as a means of identifying and managing the significant opportunities and threats
- Anticipate future resource needs and the overall resource productivity of the organization
- Establish processes appropriate to achieving the organization's strategic objectives
- Establish and maintain processes for continual improvement, innovation, and learning

These processes for sustained success apply to any organization, regardless of size, type, and activity. Attention should also be paid to several critical risk management and sustainability approaches to different organizational elements.

PRACTICES FOR EMBEDDING SUSTAINABILITY IN AN ORGANIZATION

Risk management and sustainability are critical components of an organization's sustained success. These practices should be embedded in the decision-making process by including them in the knowledge, policies, organizational culture, strategies, and operations. The organization needs to build internal competency for sustainability, engage with all stakeholders, and regularly review its actions and practices using the attributes of organizational sustainability.

IMPORTANCE OF ORGANIZATION'S CONTEXT

To begin the embedding process, it is essential to make sure that the mission statement has been converted to a clear set of strategic objectives and that the cascading of the objectives from the top down is in place. The goals and action plans must also be in place to provide feedback. Once this happens, the organization will have the means for the sustainability program to be embedded in the process for meeting the strategic objectives rather than having a separate set of "sustainability goals" with supporting initiatives that are not formally coordinated with the organization's objectives.

Scanning the organization's internal and external context helps identify its opportunities and threats. Sustainability practitioners must be involved in this search for opportunities to ensure they are adequately identified. This is a crucial linkage for sustainability to make as it starts the embedding process. Scanning the external operating environment provides the best view of the organization's social, environmental, and economic requirements. Examining the legal and other requirements would be a part of this effort. For example, some organizations have contractual commitments to follow their customers' supplier codes of conduct. There may be other contractual requirements for customers and a parent operation.

Scanning of the internal and external operating environments also helps identify the stakeholders. These individuals and organizations are associated with the opportunities and threats identified in this activity. Sustainability is usually outwardly focused. Now that the opportunities and threats are associated with uncertainty, there is an internal need for sustainability to help lower the amount of uncertainty so that the risk of meeting the strategic objectives is improved. Sustainability can ensure that the stakeholders are engaged and that there is a proper sense-making effort to determine changes in the internal and external operating environments that would dictate when a new scan is needed.

This scanning activity can be expanded within the organization's value chain. Suppliers, contractors, customers, and business partners must be engaged through sustainability to manage the uncertainty associated with these relationships. All the value chain members need to conduct their scans of the internal and external operating environments. Uncertainty can come from any point in the value chain. Part of the relationship is to provide some support to value chain partners to help lower the uncertainty of the entire value chain to promote the "upside of risk," where the opportunities are offsetting the threats to manage risk within a tolerable or even beneficial level.

It is also crucial for an organization to be aware of the level of commitment of the leaders to use sustainability, along with other organizational risk management methods, to ensure the organization is always in a comfortable position to meet its strategic objectives. Where there is an active focus on meeting organizational objectives, it is easy for the leaders to recognize the roles played by the different functional units in keeping the level of risk in line with expectations. When sustainability is seen as an active participant in managing uncertainty and risk, it is much easier to obtain the leadership's approval. The contribution of sustainability to meeting strategic objectives becomes apparent when embedded in these processes.

UNDERSTANDING ORGANIZATIONAL SUSTAINABILITY

A key role for sustainability in an organization is participating in due diligence activities. This is a process to identify the actual and potential positive and negative social, environmental, and economic impacts of an organization's decisions and actions. This happens with the active management of the internal and external context. It is no longer the case of just looking to avoid and mitigate the adverse effects (threats), but it is more about using the positive effects (opportunities) to create a positive upside of risk. This due diligence also involves influencing the risk and uncertainty analysis conducted by other organizations in the value chain. With the speed of news in this digital media world, much negative publicity can happen suddenly when a supplier encounters a problem. Therefore, increased communication and cooperation within the value chain are now seen as critical to the long-term success of any organization of any size or type.[2]

When the organization conducts its uncertainty analysis, it identifies the opportunities and threats that are most significant to the organization. This is a formal process that can be either qualitative or quantitative. An uncertainty (i.e., risk) map identifies the organization's opportunities and threats. Stakeholders need to be involved in this process. If the engagement process works well, the stakeholders will be a valuable source of information for the organization. They can help alert the organization to changes in the external operating environment and the need for an update of the uncertainty analysis. In addition, they can help in the offsetting of the threats with the opportunities that have been identified.

In the areas of sustainability and corporate social responsibility, this is often referred to as a "materiality" determination. However, these determinations rarely use uncertainty analysis and tie the results to how risk is managed within the organization and its value chain. So, do not confuse the formal uncertainty analysis with a materiality determination.

One area that sustainability programs find difficult is the determination that the organization complies with local, state, and federal laws and regulations governing its operations. A careful review of the opportunities and threats often indicates that some nuisance areas involve odors, vibrations, light, noise, and traffic, among other things. While these might meet the legal tests, the situations can irritate the stakeholders. Likewise, a product sold in a retail store might be made in a country with serious human rights issues or environmental regulations that are not enforced. These interests should be found in the scanning activities; however, a sense of complacency is associated with meeting the requirements. This is just one more reason to have a robust stakeholder engagement program. Stakeholders can help seek opportunities as well as threats.

EMBEDDING SUSTAINABILITY

Embedding sustainability into every aspect of an organization involves attention to the items covered in the elements presented in this chapter. This is much more effective than adding sustainability to a long list of expectations for members or employees. Instead, the approach to sustainability should make it a part of what everyone does every day and use its influence throughout the internal and external context of the organization.

Engaging the leaders in sustainability becomes more straightforward when it is linked to risk management. Risk and uncertainty management are almost always embedded in organizations since meeting the strategic objectives is so crucially important to any organization. Building the competency for using sustainability can become a part of the organization's drive for effective and efficient processes. Creating a culture of sustainability within the organization is more manageable when everyone sees it as a way to meet their goals so their department can meet its objectives. Meeting strategic objectives benefits everyone associated with an organization. Risk management and sustainability contribute to the ability of the organization to have an efficacious strategy.

As risk management and sustainability become embedded in the organization, there may be a need for changes in decision-making processes at all levels of the organization. There may also be some changes in governance that will recognize the positive consequences of embedding risk management and sustainability into the processes used by the organization and its partners in the value chain.

Integrating sustainability, as a separate system with separate objectives, into the organization will take a lot of time and effort, but embedding sustainability into what an organization does every day, is embraced effectively and quickly. Quality is often compared to sustainability. With the global economy and the need for continual improvement, the quality lead can be suddenly lost by replacing materials and redesigning processes to meet sustainability targets. This is much less likely to happen when all the disciplines are embedded into the processes and operations to help the organization function effectively and efficiently and maintain its social license to operate.

ENGAGEMENT, NOT SIMPLE COMMUNICATION

Communication must be an effective two-way dialogue over some time to be described as engagement. Many organizations continue to use traditional means of communication to develop awareness about the role of risk management and sustainability in helping an organization meet its strategic objectives. Frequent conversations on how the organization can improve sustainability in its processes and operations need to be interactive and acknowledged. Digital media is helping make this more effective. The dialogue must be transparent and thoughtful. It is essential to focus on helping the organization meet its responsibly set objectives over the long term. If objectives need to be changed, there needs to be a dialogue over the pros and cons of such a change and the linkage of the shift to risk management and sustainability.

Communication should be part of the processes and used in the operations. It exists to inform people and transmit important messages. Engagement takes place both within the organization and with external stakeholders. Often, an external stakeholder may start the engagement with a question, complaint, or suggestion. Then, the organization could trigger it to explain the need to make changes and invite others to participate in the decision-making process.

The people within the organization need to become familiar with the different methods and media that may be used for engagement. Thoughtful answers come from discussions with others before responding. This allows other people to join in the dialogue.

A particular form of engagement is with the media; organizations that deal with the media regularly get to know someone at the outlet and invest time to get to know them better. The building of trust will help enable some engagement to commence. The focus should always be on how the organization can meet its strategic objectives while providing shared value to external stakeholders. It makes sense to want a vibrant community and invest to see what happens when you know that employees are more likely to remain with the organization when the community is robust and healthy. In these ways, there is a difference between handling sustainability as a separate program and having sustainability embedded in the organization and made part of what every employee does every day.

REVIEWING AND IMPROVING ORGANIZATIONAL SUSTAINABILITY

Effective risk management and sustainability performance depend on the commitment, participation, engagement, evaluation, and review of contributions to effective processes, efficient operations, and efficacious strategy. Monitoring and measurement related to sustainability will come from the people engaged in using them to do their work every day. There will also be a means for collecting information on how the significant opportunities and threats were managed to enable the organization to be on the upside of risk.

Living in an uncertain world makes monitoring the external operating environment and the entire supply chain imperative. However, these numbers are not created for a report. Instead, the measurements are part of an organization's continual improvement, innovation, and learning.

Contrary to popular thought, one does not just pick processes and operations and measure them. There is a formal process of monitoring and measurement that needs to take place. It will produce information for operating the organization and engaging with external stakeholders. The data will be transparent, and the organization will be accountable for improvement to meet its strategic objectives.

ESSENTIAL QUESTIONS FOR IMPLEMENTING ORGANIZATIONAL SUSTAINABILITY

1. Why is it important to establish multiple connections between the various structural elements of a sustainability program, as discussed in this chapter?
2. How do the attributes of organizational sustainability help an organization make these connections between the structural elements?
3. How does this chapter help make it clear that structural elements are needed to define a sustainability program and embed it in how the organization operates daily?
4. Why is it essential to have a monitoring and measurement program to determine how effectively this integrated structure works?

REFERENCES

1. A. (. o. A. o. N. Zealand), *Risk Management Guidelines, Companion to AS/NZS ISO 31000:2009. HB 436*, Sydney: SAI Global Press., 2013.
2. ISO (International Organization for Standardization), *Social Responsibility Guidance. ISO 26000*, Geneva, 2010.
3. ISO (International Organization for Standardization), *Managing for the Sustained Success of an Organization: A Quality Management Approach. ISO 9004*, Geneva, 2009.

12 Implementation Is Doing

ABSTRACT

What needs to be measured? What methods should be used for monitoring, measurement, analysis, and evaluation to ensure valid results? The organization needs to evaluate the performance and effectiveness of its sustainability program. Measurements are relevant and critical to major organizational decisions and help the organization track its progress toward meeting its strategic objectives. Yet, we do not find a prominent and practical means to conduct a sustainability program's relevant monitoring and measurement tasks. Sustainability practitioners are focused on gathering available data and reporting the information to interested stakeholders in the external operating environment. Understanding monitoring and measurement is critical to the success of sustainability and is a key component in helping an organization meet its strategic objectives. Measurement also supports the organization's pledge to continual improvement.

WHY DOES AN ORGANIZATION MEASURE?

Measurement is a quantitatively expressed reduction of uncertainty based on one or more observations. The observations are made as part of the monitoring process. If something can be observed in any way, it lends to some measurement method. And those things most likely to be deemed immeasurable are often addressed by relatively simple measurement methods.[1]

Measures are needed to guide the organization to realize its strategic objectives and mission statement. At each level of the organization (i.e., strategic, operational, and tactical), a family of measures is needed to create a complete picture of that level's strengths and weaknesses so that appropriate decisions can be made with the correct follow-up actions taken.[2] Typically, the leaders are focused on the financial viability of the organization. People at the tactical level are focused on their goals and action plans. They are watched daily and measured in physical amounts: units, pounds, inches, minutes, and components. Meeting the action plans is the driving force. The middle of the organization uses screening or diagnosis measures. Is the data from the tactical level demonstrating that the tactical and operational objectives are being met? Priorities on the measures are always set according to the stakeholders' interests.

Some of the critical decisions for using measures include:[2]

- Any action, process, or operation can be measured.
- Consistent attention must be paid to the consistency of measures to determine trends.
- What is done with the measures is as important as what the measures are.

DOI: 10.1201/9781003255116-12

- Stakeholder interests should be closely related to the strategic objectives.
- Measurements must look at both lead and lag indicators. Lead indicators are the performance drivers that communicate how results (lag indicators) are to be achieved.
- There is more to organizational success than financial success.

All these decisions help an organization manage by fact. However, the "right" measure is the one that tells an organization's leader whether the stakeholder interests that are critical to the long-term success of the organization are on target.

DECIDING WHAT TO MEASURE

Organizations need to have effective systems and processes for determining what data and information should be collected and how they are handled, stored, analyzed, and interpreted.[3] This information is used to increase the organization's understanding of the internal and external operating environment within which it operates. Therefore, data and information must be continually reviewed to remain current, meaningful, and effective.

The organization should establish criteria and processes for determining what data should be collected and for not collecting data that is not useful for these purposes. Processes must have links between data gathering and the purpose for gathering the data. Data sources need to be reviewed regularly to monitor their effectiveness and continual improvement in meeting changing organizational requirements. Based on scanning the external operating environment, a wide range of stakeholders are involved in determining what data should be gathered.[3]

Organizations provide for the analysis and interpretation of the data to learn and inform its decision-making in the short and longer term. The data analysis helps the organization understand the nature and impact of variation on processes, products, services, and measurement systems. Data is also used to maintain awareness among internal stakeholders and create incentives for innovation.[3]

An organization must establish systems to ensure that data is shared among those who need to use it to improve performance and that the data is generally accessible. Data collection needs and methods must be reviewed regularly. It is also essential to check the sampling methods and sampling systems.[3] These responsibilities help ensure that the data is valid, reliable, relevant, timely, secure, and sound.[3]

Sustainability managers need to figure out what sustainability should do when embedded within an organization. What does sustainability mean to processes and operations, and why does it matter? To many people, sustainability is a vague concept that needs to be associated with what is expected to be observed. However, once everyone knows the answers to these questions, the situation starts to look more measurable.[1] The decision-making for sustainability would progress as follows:[1]

- If sustainability matters, it is likely to be detectable and observable.
- If it is detectable, it can be detected as an amount or range of amounts.
- If it can be detected as a range of possible amounts, it can be measured.

It is also important to state why it is vital to measure sustainability to understand what is being measured. Understanding the purpose of the measurement is often the key to defining what the measurement is supposed to be. Some things seem "intangible" because people have not defined them using words everyone understands.[1]

MEASUREMENT METHODS

Measurement involves using different types of sampling and experimental controls. This involves taking small random samples of some activity. It is possible to learn what to measure by measuring the situation of interest. It is important to get past the point where decision-makers seek to avoid making observations by offering a variety of objections to the measurements. To make some productive progress, consider four practical measurement assumptions:[1]

1. The problem is more common than previously thought.
2. There is more data than initially thought.
3. Less data is needed than initially thought.
4. An adequate amount of new data is more accessible than previously thought.

Think of measurement as iterative. First, start measuring what sustainability is contributing to the organization. You can always adjust the method based on the initial findings. Making a few more observations can help you learn more about what you are trying to measure and for what reason.

CHARACTERISTICS OF A MEASUREMENT SYSTEM

Even though there is much disagreement, the prevailing thought on the maximum number of metrics for an organization is around 20. Think about it. How many things can be monitored and controlled regularly? This is why key performance indicators (KPIs) were invented. Some larger organizations take thousands of metrics and roll them up to a small number of KPIs. Organizations are always advised to see their metrics like an automobile dashboard. It has a few measurements that need to be monitored regularly and a few that can be observed with a lower frequency. Those metrics not vital to meeting the organization's strategic objectives can serve as "warning lights" for leaders.[4]

There is much interest in the topic of sustainability results, but there is less written on how the results must be linked to help the organization meet its strategic objectives:

> Your measurements must focus on the past, present, and future, and be based on the needs of the customers, stakeholders, and employees. Measuring everything is more damaging than measuring nothing—pinpointing the vital few key measures is the key to success.[4]

An organization needs to evaluate its current approach to measurement and then redesign the monitoring, measurement, analysis, and review process. Having insufficient data is worse than having no data at all.

All organizations collect some types of data. Trying to track everything for everyone is a big problem in sustainability. It is crucial to locate and measure the correct variables. But it is more important to monitor and measure the organization's processes correctly. All KPIs should be linked to the organization's success factors to help meet the strategic objectives. These factors may be core values, technical competence of the workers, or marketing success. These links will help the leaders know how close the organization is to its strategic objectives.[4] Sustainability results must be linked to activities, processes, systems, products, and services associated with the organization's operation.

An organization needs to have effective systems and processes for determining what data and information should be collected. There is also a need for efficient and effective processes to acquire, analyze, apply, and manage information and knowledge. These measurements provide critical data about key processes, outputs, results, and other organizational outcomes. There is a need for information that is internal to the organization. The core practices include many of the following, whether formal or informal:[3]

- Planning of data collection and linking it to the strategic planning
- Analyzing and interpreting data to learn and inform strategic planning
- Sharing the data among those who can use it to improve performance
- Ensuring data integrity (i.e., valid, relevant, timely, secure, and sound)

Some organizations use measures from the past. For example, how did the costs today compare to the exact location's costs a year ago? We do this when we look at our electricity bills. Measuring the current performance is critical to any organization, even though the moment it is measured, that result is in the past (i.e., lag measures). Past and present metrics are the easiest to come up with because we typically have data on these measures, and these results have happened. Therefore, it is essential to have lead measures in the measurement system.

Selecting the right metrics or measures is much more than deciding what to measure. It is a vital part of the organization's strategy for success. Choosing the wrong metrics will put a strain on the success of the organization. Many small businesses are using a generic "balanced scorecard" method to have fewer measures and still be able to drive success. Organizations are learning that success is about balance, not a dependence on any fixed set of standards.[4]

Like objectives, metrics need to be set at the highest levels of a hierarchical organization and followed down to all levels and functions. Metrics at one level should lead to metrics at the next level of the organization. Defining key performance measures in this manner ensures no disconnects or inconsistencies in how the parent organization measures performance.

One way of reducing the number of measures to a reasonable number is to assign a weight to each measure in a family of metrics and develop an "index" that is an aggregate statistic. For example, several national performance measurement programs enable organizations to score their results and then aggregate them to create an aggregate category (e.g., environmental impact). Combining multiple metrics into a single index is an excellent way of aggregating and simplifying reporting on performance.[4]

STRATEGIC MEASUREMENT MODEL

While measurement may be easy, it isn't easy to measure the right things and learn to ignore other exciting data that does not help the organization become more sustainable. There have been several approaches to selecting and using metrics in an organization. Larger organizations and parent organizations use externally derived collections of sustainability metrics to guide the development of their strategy[5] as shown in Figure 12.1.

The argument is made that sustainability reporting is a form of internal monitoring and management to engage stakeholders effectively. With the new ESG requirements on the horizon, sustainability reporting and the measurement that drives it have taken a starring role in the organization's strategy and how it decides what to target. This path has an organization focus on its impacts so that it can improve its operations in a manner that lessens those impacts. The organization's operations could be conducted with stewardship in mind while creating the measurements for employee engagement and ultimately meeting their reporting requirements. Therefore, it is essential to see the sustainability report as something other than the sustainability program but instead as a means of keeping track of meeting the strategic objectives over time.

It is possible to take a more internal view by looking at metrics from the internal perspective[4] (Figure 12.2). In this model, the organization determines what it stands for and its vision of the future. Its external impacts and stakeholder engagement

FIGURE 12.1 Sustainability reporting as a management process.[5]

FIGURE 12.2   Strategic measurement model.[4]

will shape the vision. Next, the organization focuses on what it needs to do to maintain its competitive advantage and its social license to operate. This would involve understanding its internal and external context. Usually, the organization will start by determining its lead indicators and how they can be measured to assess how effectively the lead indicators are working to influence the sustainability results (i.e., lag indicators). It is important to remember that results are merely the outcomes of performance. Results are not a direct measure of performance. It is important to create results that keep score so that changes can be made in the support activity that is the focus of the lead indicators.

Objectives should be set only after there has been some experience with the determinations of the critical success factors. Often, there is a rush to set strategic objectives. This leads to objectives that are surpassed with little effort or objectives that cannot be reasonably achieved. Neither of these extremes helps drive an organization to meet its strategic objectives over the long term. Metrics are the tools used to monitor the progress in meeting strategically established objectives. This reminds us that the sustainability strategy should be prepared after the measurement system is already in place. One should never start with the strategy and derive measures that measure the progress of that strategy.

ROLES OF MONITORING AND MEASUREMENT

When any determination is needed, there is a need for monitoring and measurement. A relationship diagram highlights the respective roles of monitoring and

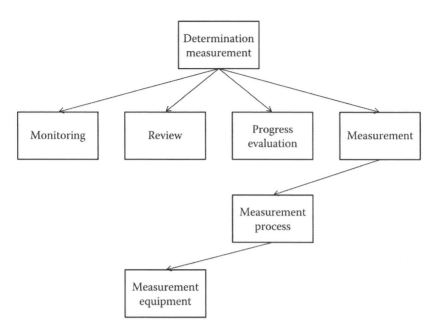

FIGURE 12.3   Measurement relationships diagram.[6]

measurement in the overall process (Figure 12.3). It is instructive to see how monitoring and measurement play different roles in the process of obtaining data for a determination:[6]

- Characteristic: Activity to find out one or more characteristics and their values
- Review: Determining the suitability, adequacy, and effectiveness of a product or service to achieve established objectives
- Monitoring: Determining the status of a system, process, or activity
- Progress evaluation: Assessment of progress made on achievement of the objectives
- Measurement: Process to determine a value
- Measurement process: Set of operations to determine the value of a quantity
- Measuring equipment: The measuring instrument, software, measurement standard, reference material, or auxiliary apparatus necessary to realize a measurable process

Note that monitoring can determine the status with observations or measurements. There are relationships between all these different practices as they work together to make essential determinations in an organization. Monitoring and measurement are not conducted for the sole purpose of addressing the interests of stakeholders. The stakeholders are interested in an outcome and want evidence of what is happening.

ESSENTIAL QUESTIONS ON MEASURING

1. Why does an organization monitor and measure to support its sustainability program?
2. How does the sustainability reporting model differ from the strategic measurement model?
3. What is the difference between monitoring and measurement? Why do you need both?

REFERENCES

1. D. Hubbard, "How to measure anything: Finding the value of 'intangibles'," *Business*, Hoboken: John Wiley & Sons, 2nd ed., 2010.
2. C. Thor, *The Measures of Success: Creating a High Performing Organization*, Essex Junction: Oliver Wight Publications, 1994.
3. Standards Australia International, "Australian business excellence framework," Sydney: Standards Australia International Ltd., 2007.
4. M. Brown, *Keeping Score: Using the Right Metrics to Drive World-Class Performance*, New York: Quality Resources, 1996.
5. International Finance Corporation (IFC) and Global Reporting Initiative (GRI), "Getting more value out of sustainability reporting," IFC and GRI, 2010. .[Online]. Available: https://www.ifc.org/wps/wcm/connect/topics_ext_content/ifc_external_corporate_site/sustainability-at-ifc/publications/publications_gpn_gettingmorevalue
6. ISO (International Organization for Standardization), *Quality Management Systems: Fundamentals and Vocabulary. ISO/DIS 9000*, Geneva, 2014.

13 Monitoring in Organizations

ABSTRACT

An organization needs effective processes and systems for determining what information and data should be collected to determine whether it is achieving its strategic objectives. Monitoring is used to determine the status of an activity, process, or system. Measurement is the process for determining a value or measure. The outputs from the monitoring are periodically reviewed to check the current situation, for changes in the work environment, conformance with organizational practices, and regulatory compliance. Monitoring, measurement, and review are activities that are used to determine the suitability and adequacy of processes and operations, and how the achievement of effectiveness and efficiency enables the organization to meet its objectives in an uncertain world.

Measurement produces results that are lag indicators. There are a number of widely used performance measurements that contain criteria that enable quantitative evaluation of organizational performance and that are applicable to the interests of stakeholders. These performance indicators are leading indicators. Their assessment criteria permit organizations to benchmark their processes, operations, products, and services with other organizations, as well as providing a useful "looking forward" sustainability perspective.

MONITORING

Monitoring involves the routine surveillance of actual organizational performance in order to create an accurate comparison with the expected or required performance. This activity involves continual checking or investigating, supervising, critically observing, and determining the process or system status to identify these changes, as well as changes in the context of the organization.[1] The main purpose of monitoring is to generate information needed to ensure that opportunities and threats are managed effectively. These effects of uncertainty and the operational controls and response may change over time. Those responsible for the embedded sustainability efforts need to be aware of the implications of any changes in the internal or external operating environment.

Monitoring requires a systematic approach that involves:[2]

- Establishing a procedure for continual checking, supervising, critically observing, or otherwise determining the status of information or systems
- Developing a means of detecting change from what has been assumed or is expected

DOI: 10.1201/9781003255116-13

- Incorporation of the organization's performance indicators and the interests of the stakeholders
- Determining how resulting information can be captured, analyzed, reported, considered, and acted upon
- Providing the necessary resources and expertise for these activities
- Allocating responsibilities for various risk management and uncertainty monitoring activities and incorporating those responsibilities in the member's or employee's performance review criteria

Monitoring is an aspect of effective governance to make sure that uncertainty is managed effectively. The selection of things to be monitored should be focused on the significant opportunities and threats.

Next, it is important to determine *what* needs to be monitored. Here are some typical monitoring examples:

- Key characteristics of operations that have significant impacts on the environment, society, or the local economy and meeting the organization's three responsibilities
- Processes in place that ensure compliance with legal and other compliance obligations
- Key characteristics associated with the interests of stakeholders
- Significant organizational opportunities and threats
- Operational controls associated with the system of management
- Value chain controls associated with suppliers, customers, and stakeholders
- Interactions found in the value chain model
- Determinants of progress toward the organization's strategic objectives

The organization needs to determine a method that will be used for monitoring that is consistent with the measurement method selected. It is also important to ensure that any monitoring device (e.g., sensor) is calibrated and maintained as required by the manufacturer's specifications.

Documented information is needed to review the results of the monitoring activity. Included in these records are the following:[3]

- Information on how monitoring decisions were made
- Information regarding the interests of stakeholders
- Controlled records for any reporting that is required
- Assistance to the organization in the review of its activities, processes, and systems
- A demonstration of whether the process has been conducted in a planned and systematic manner
- Enabling of information about the process to be shared through engagement with internal and external stakeholders
- Provision of objective evidence for internal and external process audits

Leaders need to establish and maintain processes for monitoring and collecting and managing the information obtained from the monitoring activities.

MEASUREMENT

All organizations collect different types of data. Trying to track too much information is a big problem in the sustainability arena. It is important that the organization measure the activity associated with the significant opportunities and threats, the effectiveness of the processes, and the efficiency of the activities, processes, and systems. These measurements provide critical data about key processes, outputs, results, and other outcomes. Measurement focuses on the past, present, and *future* (i.e., looking forward results) and may be based on the needs of employees, management, customers, and stakeholders. The key to a successful measurement program is pinpointing the vital few key measures.[4]

Measurement is a process for determining a value.[5] Measurements should be either quantitative or semi-quantitative. It is important that measurements be conducted under controlled conditions with appropriate processes and procedures for ensuring the validity and traceability of the results. These processes include adequate calibration and verification of monitoring and measurement equipment, use of qualified people, and use of suitable methods and quality control procedures. Measuring equipment should be calibrated or verified at specified intervals or, prior to use, against measurement standards traceable to international or national reference standards. If the standards do not exist for a particular measurement, the basis used for the calibration and use of the equipment must be recorded. Written procedures for conducting measurement and monitoring are necessary to provide consistency in measurements and enhance the reliability of the data that is produced.

Leaders need to assess the progress of the organization in achieving planned results against its strategic objectives, at all levels and in all relevant processes and functions. A measurement and analysis process is used to gather and provide the information necessary for performance evaluations and effective decision-making. The selection of appropriate key performance indicators (KPIs) and measurement methodology is critical to the success of the measurement and analysis process.[6]

LAG INDICATORS

Lag indicators are measures that focus on the outcomes or results of planned initiatives at the end of a time period. Results or KPIs largely indicate historical performance. There are large numbers of targeted sustainability indicators and KPIs available in the literature. However, the challenge in using them includes the fact that these results do not reflect current activities. They lack predictive power. There is a caution in the investment world that states, "Past performance offers no indication of future results." Past performance is useful in valuing the sustainability program only as far as it is indicative of what is to come in the future since sustainability is always defined as happening over the long term. Many sustainability practitioners regard past performance as being able to at least provide an idea of the direction in which the organization is headed. However, a performance trend is not considered to be a leading indicator.

Although sustainability practitioners and stakeholders tend to focus on results (lag indicators), these lagging indicators may not provide enough information to guide actions and ensure success.[7] There are some good reasons why lag indicators may not always be sufficient for measuring sustainability:

- Lag indicators often provide information too late to enable sufficient response.
- Outcomes are the result of many factors. Lag indicators may tell you how well you are doing but may not give information as to why something is happening and where to focus any corrective actions that may be necessary to improve performance.
- When the outcome rates are low, there may not be sufficient information in the measures to provide adequate feedback for effective management of the process.
- When the outcomes are so threatening that you cannot wait for it to happen before the process is upset.
- Lag indicators may fail to reveal latent threats that have a significant potential to result in disaster.

Lag measures tell you if you are about your success in terms of achieving the strategic objective as a "work in progress" measurement. It is difficult to do anything about a lag measure once it is reported because lag measures have already happened when you measure them. You are always referring to the past when talking about a result (lag measure). They are like a photograph—activity frozen in a past point in time. Yet despite the problem of not possessing a looking forward capability, lag measures are believed to be a sign of succeeding with a strategic objective by getting the results to demonstrate this achievement.

LEAD INDICATORS

A lead performance indicator is something that provides information that helps the user respond to changing circumstances and take actions to achieve desired outcomes or avoid unwanted outcomes. The role of a lead indicator is to help organizations improve future performance by promoting action to correct or avoid potential weaknesses without waiting for something to go wrong. The ability to guide actions to influence future performance is an important characteristic of lead indicators.

There are two forms of lead indicators. One form of a lead indicator involves providing information about the current situation that will affect future outcomes. A second form of lead indicator involves the measurement of processes involving people that can make a process more effective and connected to meeting the organization's strategic objectives.

A good example of the former lead indicator is when an economist uses certain lag indicators, such as "new housing starts" that foretell of the demand for goods necessary to build and furnish those homes. This reinforces the tendency of finding patterns with results (lag indicators).

The latter form of leading indicator involves the measures of activities connected to meeting the organization's strategic objectives. These measures are within the control of the organization. This lead measure (e.g., leadership) is different since it can foretell the result. You can recognize lead measures because of the following characteristics:[8]

- A lead measure is predictive—if the lead measure changes, you can predict that the lag measures associated with the lead measure will also change.
- A lead measure is influenceable—it can be influenced by the organization that seeks to manage the activities connected to meeting the organization's objectives.

For many people, lead measures seem to be counterintuitive. For example, we know that leaders and sustainability practitioners always focus on the lag indicators, even though it is not possible to act on a lag measure since it is always in the past. There is a feeling that lead measures are hard to keep track of since they measure behaviors and behavior change. Finally, for too many people, lead measures often look too simple. Since lead indicators focus on behaviors such as leadership, employee, and stakeholder engagement, those that are not involved with these activities directly think that they are not as important as a measure for sustainability.[8]

In the fields of organizational development and risk management, we are focused on behaviors (i.e., positive effects) that help the organization achieve its strategic objectives. The information on lead indicators makes a difference and enables the organization to close the gap between what the organization should do and what it is actually doing. Risk managers often describe lag indicators as driving an automobile with the front windshield painted black, moving forward by navigating with the three rear-view mirrors. So how does an organization prepare for creating lead indicators to drive the lag indicators over the long term?

LEAD INDICATORS AND THE PROCESS

Processes always present some challenges since people are focused on results:[8]

- Is the process producing good results?
- Are people carefully following the process?
- Is the process effective with efficient operations?

The process focus element of performance frameworks looks for leverage points in the process. These are those critical steps in the process where performance can falter. Lead indicators are linked to these points to improve the results using leadership,

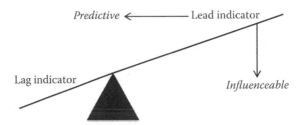

FIGURE 13.1   Acting on lead indicators.[8]

strategic planning, employee engagement, stakeholder engagement, information and knowledge management, and process focus.

The trick is to use the leverage of the people support processes in the Porter value chain model to trigger the improvement of the results (lag measures) over time. The approach, deployment, assessment, and refinement (ADAR) approach is used to accomplish this. Once you have a goal to meet an objective, there is a need to look to the performance area that would be responsible for providing the leverage factor, as shown in Figure 13.1. The process focus category in the performance framework can be used to guide the development of the approach. This is followed by a plan to deploy the process focus approach to guide the improvement of the lead measure. When this combination works well, the lead measure will be predictive and influence the lag measure. The lag measures that are attached to each lead measure should vary in the same manner.

PERFORMANCE FRAMEWORKS

Performance frameworks have been used in approximately 70 countries. These frameworks provide the foundation for what are referred to as national quality award programs. Performance excellence programs were introduced in the late 1980s to spur the competitive advantage of businesses in those countries. In the United States, the Baldrige Performance Excellence Program has been in continuous use since 1987. This program has developed a set of criteria for performance excellence to be used by participating organizations, which include large and small businesses, government, nonprofits, schools, and healthcare facilities. There is a scoring method that enables the assessors to score the applications and grant the award to those who perform best.

Many organizations have learned to use the performance frameworks as a means to improve their competitiveness without applying for the award. Because the frameworks are updated on a regular basis to reflect changes in the best practice for each of the categories, they provide a useful means for looking forward and driving performance.[9] This, in turn, helps improve the results. The developers of these frameworks noted that results are merely the outcome of performance and do not measure performance directly. What is different with the performance frameworks is their focus on driving performance in each of the following areas:

- Leadership
- Strategic planning
- People and employee engagement
- External stakeholder engagement (including customers)
- Information and knowledge management
- Process focus

Each of these areas can be scored and act as a lead indicator. Each lead indicator is associated with results (lag indicators) to see if the lead indicators are influencing and predicting the lag indicators. By measuring the inputs to a process, lead indicators can complement the use of results (lag indicators) and compensate for some of their shortcomings. Lead indicators of this nature can also be used to monitor the effectiveness of process controls giving advance warning of any developing weaknesses before problems occur.

USING A PERFORMANCE FRAMEWORK

A performance framework contains information on what is recognized as a best practice in each of the categories.[10] To keep current, the framework must be continually modified to reflect current best practice. Any organization can use these descriptions to benchmark against and make corresponding improvements in its processes and operations.

Three documents are prepared by the organization for each of the performance categories selected to help guide the efforts to attain the level of performance that is sought. These documents include an approach, the means of deploying that approach, and the manner in which the organization will assess and refine the approach and the deployment of the course of the implementation period—usually one year.[11]

A description of how to prepare each of these documents is provided below. This is followed by a description on how to score them so they can act as a semi-quantitative measurement. These measurements will become the lead indicators for the sustainability program.

APPROACH

To create semi-quantitative lead indicators that drive performance for each of the sustainability indicators, an organization needs to create its strategy. How will it address the various human support processes listed above to improve its current operating state? The approach of the ADAR cycle is basically the planning stage of the sustainability process. An organization must establish the strategic and operational objectives and the processes necessary to deliver sustainability results or KPIs.

The approach identifies the organization's intent to become sustainable and to measure its progress. In other words, the approach is the way by which something is made to happen.[11] It consists of processes and actions that take place within a framework of principles and policies.

In order to create the approach for each category, it is important to answer the following questions:[12]

- What is the organization seeking to achieve for each category? What is its intent?
- What strategic and operational objectives and worker goals and action plans have been established?
- What strategies, governance, and processes have been developed to achieve the intent? How were these items chosen by the leadership of the organization?
- What indicators or KPIs have been designed to track the progress of the organization's performance?
- How does the approach provided align with the strategic objectives established by the organization?

The approach should not only specify what an organization is planning to accomplish, but also include the reason why it is planning to do this.[11] In order for the approach to be sound, the organization should consider many of the following:[13]

- Interests of the stakeholders as determined in the engagement process
- Alignment with the organization's strategic objectives and strategy
- Appropriateness for the organization's internal and external context
- Links to other approaches, if appropriate
- Sustainability principles
- Whether it enables the organization's approach to continual improvement, innovation, and learning
- Consideration for the scoring matrix used to create the semi-quantitative score

By addressing the information provided above, the approach will ensure that an organization's sustainability strategy correlates with a planned and strategic cycle of improvement. The stronger the adherence to these items, the higher the score that will be given to the approach. When items are not addressed affirmatively, the score will be lower. A weak approach typically addresses what was to be accomplished, not how or why. A well-defined approach that is supported with some benchmarking information tends to score higher using the scoring matrix.

DEPLOYMENT

The deployment of the approach identifies how an organization plans to implement those processes and activities to create this lead indicator component. Successful deployment involves embedding the approach into the activity and full acceptance of the process steps and activities found in the approach. The deployment goal is to achieve the intent that was stated in the approach.

To successfully implement the approach, the organization should consider preparing a draft action plan.[14] Essentially, the action plan heading should describe the area being addressed (i.e., leadership, strategic planning, employee engagement, and so forth). It should also include the purpose of the action or the intent and state the expected benefits. Next, each of the tasks should be outlined assigning them to individuals, providing a timeline, and highlighting the anticipated performance improvements required for completing the item. The tasks will be the key process steps and procedures outlined in the approach. Usually, the first task of the action plan should describe what is being measured during the deployment and the last step should recap the lessons learned by the employees and leaders as a result of the deployment of the approach.

To assess the success of the deployment (i.e., implementation plan), organizations should be able to describe the following:

- How have the strategies, processes, and activities been executed with the employees? What is the effectiveness of their implementation throughout the organization? To what extent have they been accepted and embedded as part of normal operations?
- Has the approach been implemented across all relevant organizational areas to its full potential and capability? Do the planned implementation activities support operations and sustainability improvement?
- Has the deployment of the approach been executed in a timely fashion and structured in a way that enables it to adapt to changes in the internal and external operating environment?
- Is the deployment of the approach achieving the planned benefits?
- Is the deployment of the approach understood and accepted by the internal and external stakeholders?

A written action plan will facilitate the evaluation of the deployment and enable the organization to address each of the questions that are posed during the scoring of each approach and deployment.

ASSESSMENT AND REFINEMENT

Assessment and refinement provide a process of improvement that enables the approach and deployment statements to be reviewed and modified to achieve the best possible results. Through monitoring and analysis of these results, the level of effectiveness of the sustainability program can be measured. An efficient and highly rated organization will conduct learning activities in the organization in order to prioritize, plan, and promote creativity, innovation, and improvement. This leads to three categories of an assessment and refinement: measurement, learning and creativity, and innovation and improvement.

MEASUREMENT

The approach and deployment must be regularly monitored and measured to ensure efficiency and the implementation of processes. Monitoring and measurement enable the organization to indicate that the results indeed stem from the deployed approach. The organization should be able to answer the following questions to assist in the assessment and refinement:

- What processes are in place to review the appropriateness and effectiveness of the approach and deployment?
- How do the results of the approach and deployment compare with benchmarks or previous levels of performance?
- How do you communicate, interpret, and use these results?

LEARNING AND CREATIVITY

In a learning stage, an organization identifies the best internal and external practices so that it will be able to recognize improvement opportunities in its sustainability program. Through the use of creative measures, the organization can then refine the approach and deployment by encouraging continual improvement and innovation. Answers to the following questions are addressed by learning and creativity:

- What have you learned?
- How have you captured this learning?
- How can you use the learning to improve the approach and deployment?

IMPROVEMENT

Finally, the organization should use the results from the measurement, learning, and creativity to plan and implement the identified improvements. These refinements and innovations should be shared across all relevant approaches for consistency and continual improvement. At this point in the process, the organization can state how it has used the learning to improve the approach and deployment.

SCORING THE ADAR CATEGORIES

Performance frameworks score the ADAR categories using a special scoring matrix (Table 13.1). The approach is analyzed to determine its soundness in providing a clear rationale, defined processes, and a focus on stakeholder interests. It should have strong ties to the sustainability and the organization's efficacious strategy. The deployment is analyzed to determine whether it is systematic and effectively implemented. It is important to ask whether it was conducted on time and in a structured way. The scoring of the approach and refinement is threefold, beginning with the measuring of the effectiveness and efficiency of the approach and deployment, and then learning and creativity, and improvement and innovation.

When scoring each ADAR category, the assessor should start by reading the information in the middle row (i.e., 5 points). If the evaluation is deemed to be better than that description, the assessor looks at the low and high side of the next higher row (i.e., 6–7 points). On the other hand, if the evaluation is deemed to be less

TABLE 13.1

Performance Scoring Matrix

Score	Approach	Deployment	Assessment and Refinement
0	No evidence that the approach has been considered and there is a reactive attitude to performance.	Anecdotal information on how the approach has been deployed. No evidence that a deployment action plan has been used.	Anecdotal information on how the approach and deployment have been addressed. No results shown and no improvement activities in place.
1–2			
3–4			
5	Approach has a rationale and is proactive with defined processes. May not address strategy and recognition of stakeholder interests. No improvement from assessment and refinement.	Approach is applied to many areas and activities. Approach is becoming part of operations and planning with formal action plans and accountability. Evidence that results are caused by approach in some areas.	Positive trends in many areas. Results are comparable with benchmarked findings in many areas. Evidence that results are caused by the approach and deployment is subject to ad hoc review. Evidence that some improvements are implemented.
6–7			
8–9			
10	Approach is accepted as best practice and is benchmarked by other organizations seeking to improve. All items in the guidance are evident.	Approach is deployed to all areas and activities. Approach is totally integrated into normal operations and planning.	Results are clearly caused by the approach in all areas. There is a proactive system for regular review and improvement of the approach and deployment.

Source: Adapted from EFQM, EFQM excellence model, EFQM, Brussels, 2010.

positive than the middle row, the assessor looks to the wording in the next lower row (i.e., 3–4 points). The movement continues until the assessor is comfortable with the evaluation in the scoring matrix as a reasonable explanation of what was observed. The assessor scores each of the ADAR categories in order, from right to left in the scoring matrix. Typically, the deployment lags the approach and the score is typically at least 1 point lower. The assessment and refinement lags the approach and the deployment, so the score is typically 1 point lower than the deployment score. It takes a little practice to use the scoring matrix.

Remember that the leadership is using the scoring sheet when preparing the approach, deployment, assessment, and refinement documents. Some organizations start with a modest approach in the range of 5 points. If successfully put into use, the approach will generally score between 4 and 6 points. The scores get higher when the organization has more experience with this process. As mentioned, there are typically up to six processes that are implemented as leading indicators, as found in the top section of the Porter value chain diagram. No organization is equally proficient in the ADAR of the six lead indicators. You may see some approaches in the 8–9 scoring range and some in the 4–5 scoring range. As the best practices improve, it becomes more difficult to get the same score as the year before with the same amount of effort. On the contrary, the organization must improve its ADAR methods to maintain its score as a result of the improvement in best practice.

ESSENTIAL QUESTIONS ON MONITORING IN THE ORGANIZATION

1. How do monitoring and measurement work together to help an organization create and maintain effective processes and efficient operations?
2. How do you define the way lagging and leading indicators work together within the monitoring and measurement system of an organization?
3. How can you best describe each of the people activities at the top of the value chain model for a particular organization in advance of scoring each activity using the ADAR method?
4. How do the ADAR scoring categories enable an organization to quantify the effectiveness of the categories at the top of the value chain model?

REFERENCES

1. Geneva, *Risk Management: Guidance for the Implementation of ISO 31000. ISO/TR 31004*, Geneva: ISO (International Organization for Standardization), 2013.
2. A. (. o. A. o. N. Zealand), *Risk Management Guidelines, Companion to AS/NZS ISO 31000:2009. HB 436*, Sydney: SAI Global Press, 2013.
3. Geneva, *Risk Management: Principles and Guidelines. ISO 31000*, Geneva: ISO (International Organization for Standardization), 2009.
4. M. Brown, *Keeping Score: Using the Right Metrics to Drive World-Class Performance*, New York: Quality Resources, 1996.
5. Geneva, *Guidance on the Concept and Use of the Process Approach for Management Systems. ISO/TC 176/SC2/N 544R3*, Vols. Retrieved June 22, 2015, Geneva: ISO (International Organization for Standardization), 2008.
6. Geneva, *Managing for the sustained success of an organization: A quality management approach. ISO 9004*, Geneva: ISO (International Organization for Standardization), 2009.
7. S. C. i. S., "Leading performance indicators: Guidance for effective use," *Step Change in Safety*, Aberdeen, UK, [Online]. Available: https://www.stepchangeinsafety.net/node /2667.
8. C. C. S. a. H. J. McChesney, *The 4 Disciplines of Execution*, London: Simon & Schuster, 2012.
9. R. Pojasek, "A framework for business sustainability. 17," *Environmental Quality Management*, vol. 2, p. 81–88, 2007.
10. R. a. H. J. Pojasek, "Improving sustainability results with performance frameworks," *Environmental Quality Management*, vol. 20, no. 3, p. 81–96, 2011.

11. EFQM, "EFQM (European Foundation for Quality Management)," *EFQM Excellence Model*, EFQM, 2010.
12. S. Global, *Australian Business Excellence Framework*, Sydney: SAI Global, 2007.
13. E. (. F. f. Q. Management), "EFQM framework for corporate social responsibility," *EFQM*, 2003.
14. R. Pojasek, "Implementing your pollution prevention alternatives," *Pollution Prevention Review*, vol. 7, no. 2, p. 83–88, 1997.

14 Sustainability Frameworks and ESG Reporting

ABSTRACT

At the organizational level, information and stories are transmitted to the parent company for inclusion in a sustainability report. When operating separately, the monitoring and measurement activities that take place at the organizational level let the leader know whether there are effective processes and efficient operations. This data and information are shared within the stakeholder engagement program or with customers who request information.

No matter where the sustainability program is located, attention must be paid to transparency and accountability. This is increasingly important as new environmental, social, and governance (ESG) standards are being reworked globally as of the writing of this chapter. In any organization, transparency provides a sense of openness about decisions and activities that affect the environment, the community, and the economy, as well as a willingness to communicate this information in a straightforward, accurate, timely, honest, and complete manner.[1] Accountability is about being responsible for decisions and activities to the organization's governance, legal authorities, financial authorities, and, more broadly, the stakeholders.[1] The organization's performance is measured relative to its strategic, operational, and tactical objectives. Organizations seek levels of performance that deliver ever-improving value to all stakeholders, thereby contributing to organizational sustainability.[2] The quest is always for effective processes, efficient operations, and efficacious strategy. Having examined the monitoring and measurement tasks, we will now explore some of the outcomes of these efforts and perhaps discover even more sustainability measures or disclosures that need to be made to different stakeholders.

Transparency

In the practice of sustainability, an organization is expected to be transparent in its decisions and activities that impact the environment, society, or the economy. As such, a local organization is expected to disclose in a clear, accurate, and complete manner, and to a reasonable and sufficient degree, the policies, decisions, and activities for which it is responsible. In addition, this disclosure should include information on the known and likely impacts that the organization may have on the local environment, the people in the community, and its shared values.

Until the organization has secured its "social license to operate," stakeholders may expect the organization's information to be readily available, directly accessible,

DOI: 10.1201/9781003255116-14

and understandable to those who have been or may be affected in significant ways. Therefore, information should be timely and factual and presented clearly and objectively. This should enable the outside stakeholders and the organization to discuss the impact of the organization's decisions and activities on their respective interests.[1] Under the best conditions, much of the disclosure occurs within an established stakeholder engagement process.

The principle of transparency does not require that proprietary information be included in the disclosures. Furthermore, transparency does not involve providing privileged information or would breach legal, commercial, security, or personal privacy obligations.[1] Often, there is a contentious time in the stakeholder engagement process when the organization needs to build a sufficient level of trust in the view of the stakeholders. The information determines the organization's performance until trust is established.

Many organizations find themselves in a situation where the more information they provide to the stakeholders; the more information the stakeholders may demand to have provided. This can be remedied by using a risk assessment method to examine the failure modes (i.e., the impacts) and find a way to address the root cause of the situation referred to as the "impact" so that the stakeholder engagement process can focus more on whether the organization is meeting its strategic objectives and whether these objectives will have positive effects for the stakeholders as well. Therefore, failure mode effects and analysis have been adapted to manage all three sustainability responsibilities.[3]

There are several other things an organization should consider when seeking to become more transparent:[1]

- Explaining the purpose, nature, and location of its activities at the community level
- Identifying people with a controlling interest that are involved in the organization
- Manner in which decisions are made, implemented, and reviewed, including the definition of the roles, responsibilities, accountabilities, and authorities across the different functions in the organization
- Standards and criteria against which the organization evaluates its performance relating to sustainability
- Performance on relevant and significant interests of the stakeholders in the area of sustainability
- Sources, amounts, and application of its funds
- Known and likely impacts of its decisions and activities on its stakeholders, society, the economy, and the environment
- Stakeholders and the criteria and procedures used to identify, select, and engage them

Such an extensive list would be problematic for most organizations at the community level. For large hierarchical organizations, people would have to be hired to fulfill the demand for transparency in a sustainability program.

ACCOUNTABILITY

In the practice of sustainability, an organization is expected to be accountable for its impacts on the environment, society, and the economy. This suggests that an organization should accept appropriate scrutiny and have a duty to respond to this scrutiny.[1] Sometimes this concept moves away from engagement with stakeholders and assumes that watchdog organizations will be looking to see what organizations are creating impacts. Is this definition more consistent with the transparency definition if the accountability were to engage stakeholders directly and discuss the information provided transparently?

It is well accepted that accountability involves an obligation of the organization to be answerable to legal authorities about laws and regulations. However, including sustainability expands the concept of accountability from solely including an organization's overall decisions and activities on the environment, society, and the economy to those affected by its decisions and actions and the community in general.[1] Accountability also includes accepting responsibility where wrongdoing has occurred, taking the appropriate measures to remedy the misconduct, and taking action to prevent it from happening again.[1]

Organizations should base their behavior on standards, guidelines, or rules of conduct that follow accepted principles of doing the right thing or practicing good conduct in the context of specific situations, even when these situations are challenging.[1] Finally, organizations are expected to respect, consider, and respond to the interests of their stakeholders.[1] The engagement process helps establish an understanding where both parties would be expected to do the same. This is one of the outcomes when the organization is granted its social license to operate. However, we have a long way to go in how these fundamental principles get parties to engage with each other, just as the word is meant to convey.

ORGANIZATIONAL BASIS FOR DISCLOSURES

An organization seeks to meet its strategic objectives in an uncertain world. Let us use a concept often used for businesses to examine how organizations seek to disclose information on their actions to achieve these objectives. There is a continuum in all organizations that moves from the mission to strategic objectives, strategy, execution, and the determination of the organization's performance. Each organization should pay attention to the "voice of the customer and stakeholder" at each step. The outcomes of the well-executed continuum include engaged stakeholders (internal and external to the organization), effective processes, and efficient operations. Within a sustainability program, internal and external performance is usually disclosed both within the organization and externally to the organization. If there is an emphasis on the engagement of stakeholders throughout the continuum, the disclosure of performance is generally not found to be contentious. It is a statement on how well the organization and stakeholders are engaged.

The financial disclosures come from the realization of the first principle of sustainability management:

> Our organization exists to create and protect value for our members, employees, customers, and stakeholders.

Sustainability contributes to the demonstrable achievement of objectives and the improvement of organizational performance.

The society perspective represents the successful realization of stakeholder engagement—to both the internal organization and the external stakeholders and society in general. There are different levels of engagement that the organization must navigate in this perspective:[4]

- Identify stakeholders during the scanning of the internal and external operating environments
- Begin communicating with the stakeholders
- Seek engagement with the stakeholders
- Grow relationships with stakeholders

Every organization has a broad range of stakeholders.

The organization's operations can be divided into several processes that are important from the environmental perspective:[4]

- Develop and sustain the organization's niche in crucial value chains
- Produce products and deliver services in an efficient manner
- Distribute and deliver products and services to stakeholders (e.g., customers)
- Manage the organization's risks associated with these processes

The process and activity of the organization are focused on the effective and efficient conservation of resources used in services and products that can pose threats to the environment if the organization does not manage its operations, customers, and stakeholders, and three responsibilities and take advantage of innovation associated with its performance improvement.

The learning and growth perspective is driven by the sustainability management principles applied to the sustainability program. Many value chain illustrations benefit from the judicious application of these sustainability principles.

CONTEXT AND PERFORMANCE MEASUREMENT

When an organization has a sustainability program with an inward-focused perspective, it deals with stewardship of the three sustainability responsibilities: value chain management, quality management, availability of people that can accept the organization's culture, and a management system. There is a focus on efficiency, growth, financial structure, and other operational attributes. Much of the monitoring and measurement is focused on meeting its strategic objectives over the long term and successful engagement with the stakeholders. Sustainability can be embedded in the way the organization operates every day.

When an organization has a sustainability program with an outward-focused per-spective, it begins to demonstrate its awareness of how environmental, social, and economic factors affect the organization's operation. First, there is a focus on meet-ing objectives by managing uncertainty's positive and negative effects (i.e., oppor-tunities and threats). Next, it will quantify the threats and opportunities related to environmental, social, and economic factors. Finally, the organization will quantify the threats and opportunities related to its three sustainability responsibilities. This begins to demonstrate through stakeholder engagement that sustainability is part of the internal strategy and will lead to performance that recognizes both the internal and external context.

Trust is what sustainability disclosure is all about. Transparency and account-ability help the organization establish credibility. They demonstrate the effectiveness of stakeholder engagement with internal and external organizations. The organiza-tion begins to see transparency as allowing the will of the stakeholders to play its part in monitoring the sustainability disclosures' trustworthiness (i.e., validity). Organizations with an internal-only focus are much less likely to share information from the management of the internal context. The major disconnect between the three-responsibility performance and internally focused performance is the latter's focus on impacts and regulations, which are seen as a cost center instead of a value center. Many sustainability measures currently in use need to be designed to opti-mize three-responsibility performance or to understand the long-term impacts of organizational decisions. Most of the metrics currently in use focus on one responsi-bility and are internally focused.

RISK AND PERFORMANCE MEASUREMENT

Organizational sustainability practitioners use a sustainability management method called "materiality." This financial term is being applied to all the organization's nonfinancial metrics. The materiality matrices in use today have the external context (interest to the stakeholders) on the y-axis and the internal context (interest in the organization and its operations) on the x-axis.

In the materiality determination, the y-axis on the grid has one of the following labels:

- Importance to external stakeholders
- Community, neighborhood, or society concern

These concerns constitute the external context of the organization.

The x-axis or the materiality plot consists of one of the following different labels:

- Likelihood of an impact on an organization
- Influence on the organization's objectives
- Current or potential impact
- Significance to the organization

These concerns constitute the internal context of the operation.

To engage with stakeholders and improve the operation, it makes more sense to discuss the organization's uncertainty (both opportunities and threats) and how uncertainty is being addressed through the uncertainty assessment and response efforts. The metrics would include the lead indicators to show how management is dealing with the effects of uncertainty through the efforts of the people support activities found in the top half of the value chain model. The lag indicators or the results of those efforts are reflected in the suppliers, inputs, process, outputs, and customers (SIPOC) diagram at the bottom half of the value chain model. Just like the monitoring and measurement done with the process, it is best to select a few critical areas to focus on. Using the uncertainty assessment, an organization can talk about the approach to identify the effects of uncertainty and then show how it plans to manage the significant opportunities and threats, along with the risk of achieving its strategic objective.

This is a more effective approach to sharing sustainability information within the engagement of stakeholder's process as long as the organization examines all its significant opportunities and threats rather than selecting from an established list of sustainability results (i.e., lag indicators) and putting them on a materiality plot. The organization's reporting should clearly show how dealing with the effects of uncertainty can help improve its ability to meet its strategic objectives. Focusing on the strategic objectives makes it possible to manage the organization's expectations and drive the optimization of the results needed to demonstrate progress. Managing the results will involve balancing the response to the internal and external contexts by dealing with the significant opportunities and threats.

MEASURING THE LAG INDICATORS

All organizations analyze and evaluate appropriate data and information from monitoring and measurement activities. These results (lag indicators) are used to assess many of the following:[5]

- Conformity of an organization's products and services
- Degree of customer satisfaction
- Performance and effectiveness of the system of management
- Degree to which the strategy is efficacious
- Effectiveness of actions taken to address the effects of uncertainty
- Performance of the value chain suppliers
- Need for improvement, innovation, and learning

Organizations may receive requests for specific results (i.e., lag indicators) from parent organizations as part of the sustainability reporting effort or from suppliers, banks, and other organizations that are dealt with regularly. If the requested results still need to be collected as part of the monitoring and measurement effort, it is helpful to engage the parties in a discussion of the process used to gather these results. The organization should seek to:

- Measure aspects of the value chain process that create value
- Measure the activity outcomes that set the organization apart (i.e., competitive advantage)
- Measure what other similar organizations measure since this ensures that focused metrics are harmonized

However, the measures need to be mindful of stating the creation of value. The methods used by the organization to monitor and measure, analyze, and evaluate these results should ensure that:[6]

- The timing of monitoring and measurement is coordinated with the need for analysis and evaluation of the results
- The results of monitoring and measurement are reliable, reproducible, and traceable
- The analysis and evaluation are reliable and reproducible, enabling the organization to report trends

Each of the lag indicators can be judged for significance based on the following:[2]

- Is the result essential or significant to the organization?
- Is the result associated with an action plan already in use?
- Is the result regularly tracked with the organization's leaders following its trends?
- Does the organization actively benchmark this result with others?

Positive answers to these questions show that this process of tracking results is improved by the attention given to the performance management process. By engaging external stakeholders in this process, it should be possible to provide them with more valuable results while not creating extra effort within the organization. The apparent exception is the information needed to measure regulatory compliance and other similar contractual obligations.

What about the accountability associated with the engagement with other external stakeholders related to a parent organization? There may be requests for results that are not routinely collected. These requests must be discussed within the engagement process since an investment will be necessary to collect that information. The external stakeholder needs to understand the entire mechanism of monitoring and measurement, while the organization needs to know whether an interaction with the external stakeholders requires such measurement.

Within the organization, goals and action plans are set in the lowest levels of the operations. It is essential to have activities where the people involved only have one or two goals. The success of the action plan is based on the ability of the people that work with a process to meet these goals. Remember that the lag indicators are the tracing measurements used to determine whether the goals are being met. It is important to keep score. The strategic objectives are achieved through completing the operational objectives. Results must be kept at each level in the operations and the organization. Since the connection between levels is created at the initiation of the scoring dashboard, it is vital that each level only views what is important to them. The organization's leaders will manage the progress for achieving the strategic objectives and communicate that information to the entire organization and external stakeholders at the appropriate time.

The results of the sustainability program can also be scored using the four attributes of lag indicators mentioned earlier. In addition, it is possible to score the 15 or 20 key performance indicators (KPIs) or other results and then aggregate them into a single lag indicator score. More information will be provided after looking at the lead indicators.

DESIGNATING THE LEAD INDICATORS

Based on the available performance frameworks, it is possible to create several leading indicators representing people's support in the top half of the value chain model. Here is an example of these lead indicators:

- The leader embraces accountability for the organization's sustainable success.
- The organization's strategic plan is instrumental in mainstreaming sustainability.
- Engaged employees embrace sustainability as part of their daily work.
- Engaged stakeholders provide value through sense-making and process improvement.
- The organization participates in partnerships when collaboration can further improve sustainability.
- A sustainable organization is created through effective processes and efficient operations.

Every organization creates a description of how these lead indicators are realized. Examples of these descriptions[7] can be found in the following sections.

LEADERSHIP

Excellent leaders ensure the organization's mission, vision, principles, values, and ethics reflect a sustainability culture that they role-model and reinforce with the internal stakeholders to enable them to contribute to the organization's strategic objectives.

Leaders should define, monitor, review, and drive the improvement of the organization's management and performance system to ensure that it addresses current and future environmental, social, and economic factors.

Leaders should also engage directly with external stakeholders in supporting society in their community by participating in capacity-building activities. In particular, they foster equal opportunity, environmental quality, education, and health and encourage well-being among community stakeholders by minimizing the adverse impacts of their products, services, systems, and processes.

Leaders ensure that the organization is flexible and manages change effectively, considering its sustainability commitments and legal, ethical, environmental, social, and economic responsibilities.

STRATEGIC PLANNING PROCESS

Organizational governance embeds sustainability into its policies, strategy, and day-to-day activities based on understanding the stakeholders' interests in their external operating environment.

The organization's strategy is based on addressing the challenges faced by its internal performance and capabilities as benchmarked against the performance of competitors and "best-in-class" organizations for their environmental, social, and environmental impacts.

The organization's strategy and supporting policies are developed, reviewed, and updated consistently with its mission, vision, and sustainability strategy.

The organization's strategy and supporting policies are reviewed within the stakeholder engagement process and communicated with community stakeholders to raise sustainability awareness. Organizational governance communicates, implements, and monitors the strategy and supporting policies.

EMPLOYEE ENGAGEMENT PROCESS

The organization manages, develops, and releases the full potential of its people at the individual, team, and organizational levels, including involving and empowering them in discussions on sustainability and related activities and planning. This helps secure support for the organizational strategy and supporting policies.

People in the organization are encouraged and supported in developing their knowledge and capabilities. They are aligned, involved, and empowered to meet the strategic objectives and the sustainability policy as part of daily activities. People must communicate effectively throughout the organization. They are rewarded, recognized, and cared for within the organization.

EXTERNAL STAKEHOLDER ENGAGEMENT PROCESS

The organization will identify its stakeholders while scanning its external operating environment and seek to engage with them to help manage uncertainty in line with its sustainability program.

The organization will seek to obtain and maintain the social license to operate the external stakeholders grant in the local community by sharing interests and building relationships.

The organization will design and tailor processes for building and managing customer and stakeholder relationships to suit markets to acquire new customers, retain existing customers, and develop new market opportunities aligned with the organization's sustainability program.

Finally, the organization will measure customer satisfaction and loyalty; compare the results with those of the competitors; use the information to improve internal processes, products, and services; and deliver increasing value for customers, markets, and other external stakeholders.

PARTNERSHIPS AND RESOURCES PROCESS

Partners and suppliers are managed for sustainable benefit by building relationships based on mutual trust, respect, and openness.

Finances are managed to secure sustainable success by developing financial strategies to support the organization's sustainability program strategy.

Buildings, assets, infrastructure, materials, and natural resources are managed sustainably.

Technology is managed by involving internal and other stakeholders in developing and deploying new technologies to maximize the benefits.

Information and knowledge support effective decision-making and sense-making and build the organization's capability to improve its sustainability program continually.

USING EFFECTIVE PROCESSES AND EFFICIENT OPERATIONS

Processes are designed and managed to optimize stakeholder value, clearly linking the outcome measures to the organization's strategic objectives.

Products and services are developed to create optimum value for the customer while considering any impact the products and services may have on environmental, social, and economic responsibilities in the sustainability program.

Products and services are produced, delivered, and managed by involving people, customers, partners, and suppliers in optimizing the effectiveness and efficiency demanded by our sustainability program.

Products and services are effectively promoted and marketed without making any claims not supported by our sustainability program.

Customer relationships are managed and enhanced by informing them of our sustainability program to build and maintain an open dialogue based on openness, transparency, and trust.

MEASURING PERFORMANCE OF THE LEAD AND LAG INDICATORS

There is a reliable way to measure the performance of an organization that has been widely used worldwide. As a result, it is often referred to as performance excellence. The way it scores the lead and lag indicators is presented below.

The organization carefully gathers the data and information on each of the lead indicators, including documented information derived from the system of management, internal audits, management reviews, stakeholder engagement, and benchmarking efforts. Typically, a team of assessors will be trained to evaluate the information and score it using the scoring matrix. Everyone on the assessment team needs to be familiar with the approach, deployment, assessment, and refinement (ADAR) to use the scoring matrix and review this data and information.[8]

If the assessors score the components of each lead indicator, the scores are added to provide a score for each lead indicator. The scores of the lead indicators can be aggregated and divided by the number of indicators to provide a single lead indicator score. The score is usually adjusted for reporting on a 1,000-point basis.

Next, the lag indicators are measured against a scoring matrix.[2] The Baldrige scores are provided in the following results categories: "Products and Processes," "Customers," "Employees, Leadership, and Governance," and "Financial and Market." The European Foundation for Quality Management (EFQM)[7] has four result categories: "People," "Customer," "Society," and "Key Performance." All the results in a category can be scored and aggregated as a score for that category. Independent KPIs without an assigned category can also be scored. Because scores have no units of measure associated with them, it is possible to aggregate all the scores and divide by the number of indicators included to produce a single score for the lag indicators, usually presented on a 1000-point basis.

The two scores can then be combined (one of the performance excellence programs[2] combines them at 60% lead indicators and 40% lag indicators based on almost 30 years of experience) to give the sustainability program a single score. This can be done on a quarterly, semi-annual, or annual basis. This single performance can then be tracked over time to measure the organization's continual improvement.

Thus far, this chapter focuses on transparency and scoring against a standard; the concepts covered are valid for an organization regardless of the scoring or standard. The Baldrige framework was used as an example because it is relatively well known and has been operating since 1987, with its first award granted in 1988. But where things stay the same, they must also change. Developing a new, coordinated ESG reporting system has taken the world by storm. This system has not been enacted ultimately or defined by the writing of this chapter; therefore, a description has been included with the link to where the reader can find the most up-to-date information on the ESG global standard development.

DEVELOPING A NEW MODEL FOR ORGANIZATIONAL SUSTAINABILITY

According to an article in *Euromoney*, a group of officials from several international organizations worked on a plan to see if there was a set of global sustainability disclosure standards that could improve sustainability reporting. They shared their thoughts with the International Financial Reporting Standards (IFRS) early in 2019 about the possibility of such a project being used by IFRS.[9]

In January 2019, the World Resources Institute (WRI) was working with the Big Four accounting firms to realize their report of a new sustainability reporting system. They presented the progress on developing that standard at the annual meeting of WRI in January 2020. During the last week of September 2020, the IFRS and the International Organization of Securities Commissions (IOSCO) decided to issue a 90-day consultation on creating a sustainability and climate change information collection program to service the capital markets. They mentioned establishing the International Sustainability Standards Board (ISSB) to manage the program that would be developed. The consultation ended on December 31, 2020. However, based on the preliminary information and the great turnout, IFRS decided to move forward with this program. In April 2021, WRI joined the IFRS program.

With a lot of hard work and the support of many companies and individuals, IFRS announced the creation of the ISSB with the objective of formulating a new baseline for sustainability disclosures to meet capital market investor needs.[10]

IOSCO joined forces with IFRS to use the knowledge they gained working together for about 20 years to help maintain an accounting system for 30 countries as they negotiated irregularities and other problems with their accounting systems on financial projects. The International Accounting Standards Board (IASB) was positioned in the IFRS hierarchy to help each other use proven methods and have some staff available in the initial phase of forming the ISSB.

On November 3, 2021, the IFRS Foundation publicly announced the creation of the ISSB and described the objective of formulating a new baseline for sustainability disclosures to meet investor requirements. The IFRS then sat alongside IASB in the IFRS hierarchy to prepare a new baseline for sustainability disclosures to meet investor requirements. The existing Climate Disclosure Standards Board (CDSB) and the Value Reporting Foundation (VRF) help manage the so-called "alphabet soup" of reporting standards bodies, as well as ensure the right expertise that came along with these deals.[11]

Obtaining good results from the consolidation was one thing, but they also needed the willingness of other bodies to merge with an organization that didn't even exist. These were challenging conditions!

This consolidation could trace its roots to nearly 15 years earlier when there was a debate over how to construct reporting that could go beyond finance and into all aspects of corporate impact and value creation. As a result, the IFRS ensured that the CDSB, a not-for-profit organization, was created during the 2007 meeting of the World Economic Forum.

The International Integrated Reporting Council (IIRC) was launched in 2010 by the Prince of Wales and changed by a former Bank of England Governor, Mervyn King. In 2011, the Sustainability Accounting Standards Board (SASB) was founded as another avenue for exploring the means to help businesses and investors develop an organized sustainability business and an organized sustainability dialogue. This was followed in 2015 by the Task Force on Climate-Related Financial Disclosures (TCFD) since it was a part of the Financial Sustainability Board (FSB). These efforts spawned initiatives that led to the criticism of "alphabet soup"—a euphemism for too many things going on and lacking clarity. This made explicit the need to make a concerted effort to move toward standardization. In addition, the CDSB, modeled on the IASB, needed a level of scale to contribute to its success. These moves helped regulatory authorities and international accounting authorities to start the process by using the CDSB as a bit of a workaround.

By 2019, the CDSB was beginning to find a way to make this work. Officials at SASB and the IIRC started meeting with CDSB to plan to create a unified reporting system. They wanted the blessing of the IFRS and the regulators that worked with the IOSCO. But at that time, there was no certainty about what would happen.[10]

In September 2020, the informal group distributed the first vision of how the pieces might fit together and widely distributed the proposal as a 90-day public consultation that attracted 600-plus responses. This showed two things:

1. There was a market demand for a global reporting solution.
2. That favored a role for the IFRS Foundation in providing it.

A technical readiness working group (i.e., World Economic Forum, VRF, TCFD, CDSB, IASB, and IOSCO) spent months figuring out exactly how this would be achieved. The team prepared prototypes, while others poured over the governance requirements and amendments to the IFRS Foundation constitution needed to move to the next step. By the time COP26 began, the pieces had fallen into place.

IOSCO worked to explain its work to its members. It was well known that if the future standard could meet IOSCO's expectations, IFRS would be endorsed to adopt, apply, or be informed by the standard.[11]

The feedback was that the new system could not replace jurisdictional authorities such as securities regulators of the policies of governments but rather be something onto which jurisdictions can add whatever they feel is appropriate. The priority was to remain demand driven. IFRS asked stakeholders if there was a need for global sustainability standards. Second, IFRS would ask if they should play a role.

At the official launch, the ISSB developed two prototypes for the standards. One is for general requirements for the disclosure of sustainability-related financial information, and the other is for addressing climate change. The climate-related prototype details nearly 70 individual areas to be reported. It has the objective of requiring entities to disclose information about exposures to climate risks and the opportunities to enable investors to form a view about future cash flows and, therefore, enterprise value.[10]

ESSENTIAL QUESTIONS FOR TRANSPARENCY, ACCOUNTABILITY, AND REPORTING

1. Why should an organization measure the contribution of the sustainability program to overall operational performance rather than simply tracking sustainability initiative progress, one KPI at a time?
2. Why would the sustainability program choose not to be recognized as an essential driver for improved performance at the facility and corporate levels by restricting its measures to lag KPIs?
3. How do the lead indicators associated with an embedded sustainability program influence and predict the lag indicators selected for disclosure by the organization?

4. How does the scoring method common to the major performance frameworks provide meaningful information for disclosure while protecting sensitive information associated with some of the sustainability initiatives?

REFERENCES

1. Geneva, *Social Responsibility Guidance. ISO 26000*, Geneva: ISO (International Organization for Standardization), 2010.
2. Baldrige, *Baldrige Performance Excellence Criteria*, Gaithersburg: National Institute for Science and Technology, 2013. [Online].
3. H. a. M. R. Duckworth, *Social Responsibility: Failure Mode Effects and Analysis*, Boca Raton: CRC Press, 2010.
4. R. a. N. D. Kaplan, *Strategy Maps: Converting Intangible Assets into Tangible Outcomes*, Boston: Harvard Business School Press, 2004.
5. Geneva, *Quality Management Systems: Requirements. ISO 9001*, Geneva: ISO (International Organization for Standardization), 2015.
6. Geneva, *Environmental Management Systems: Requirements with Guidance for Use. ISO FDIS 14001*, Geneva: ISO (International Organization for Standardization), 2014.
7. E. (. F. f. Q. Management), *EFQM Excellence Model*, Brussels: EFQM, 2010.
8. G. G. R. a. Y. J. McCarthy, "Guidelines for assessing organizational performance against the EFQM model of excellence using the radar logic," Vols. Retrieved March 25, 2016, 2002.
9. R. b. Pojasek, *Organizational Risk Management and Sustainability*, First Edition, Boca Raton: CRC Press (A Taylor & Francis Group), 2017.
10. I. F. Trustees, "Where the ISSB came from and what it will do," *Birth of a Standard*, [Online]. Available: https://www.euromoney.com/article/29a2t1n0x3f40blljhlvk/esg/birth-of-a-standard-where-the-issb-came-from-and-what-it-will-do.
11. I. Foundation, *Technical Readiness Working Group*, [Online]. Available: https://www.ifrs.org/groups/technical-readiness-working-group/#resources.

15 Organization Self-Assessment and Maturity

ABSTRACT

The maturity of an organization's management system is determined through self-assessment. Self-assessment is a comprehensive and systematic review of the organization's activities and sustainability performance concerning its degree of maturity. This process is used to determine the strengths and weaknesses of the organization in terms of managing its opportunities and threats. The self-assessment needs to cover all the internal levels of the organization, from the leaders to individuals responsible for the operations or supporting operations.

The self-assessment results should be communicated within the engagement with internal and external stakeholders. Self-assessments provide information about the organization and a perspective of its future direction. These are topics that are commonly discussed within a stakeholder engagement process. The monitoring and measurement results should be an input to the self-assessment but should not be considered a replacement for a well-designed self-assessment process. For example, stakeholders are more interested in the maturity of an organization's sustainability program than a report that discusses the sustainability results achieved by the organization in the past.

EVALUATION OF PERFORMANCE

Once the measurement results are available, leaders need to assess the organization's progress in meeting its cascaded objectives by evaluating the employee goals within the action plans prepared at the beginning of each reporting period (e.g., monthly, quarterly, or annual). There are several ways by which this evaluation is conducted.

INTERNAL AUDITING

Internal audits are commonly used in the "check" phase of an organization's plan–do–check–act (PDCA) activity. These audits help identify problems, threats, opportunities, and nonconformities with the sustainability program embedded in the operations and support operations.[1] By focusing on the management system, these internal audits help the organization monitor progress in its sustainability program and the success of closing identified nonconformities from previous audits. In this manner, an internal audit verifies that the planned sustainability actions have been practical and how these actions improve the organization's ability to meet its strategic and operational objectives.[2]

Internal audits are also helpful in identifying "good practices" that can be extended to other areas of the organization's operations, products, and services. This will help

DOI: 10.1201/9781003255116-15

the organization improve due to its sustainability program embedded within its operating system. Internal auditing is not as effective in evaluating a sustainability program restricted to working on initiatives rather than embedding sustainability into the organization's daily operations. Internal audits are an essential component of how organizations manage within their system of management.

BENCHMARKING

Benchmarking is a measurement and analysis methodology that an organization can use to search for "best practices" within its operations or by comparing activities with those of another organization. Organizations engage in benchmarking to improve the overall performance and their policies, strategies, operations, processes, products, and services.[2]

There are many different types of benchmarking:[3]

- Internal benchmarking comparing activities within the organization
- Generic benchmarking comparing strategies, processes, and operations with those of organizations that are not competitors
- Competitive benchmarking comparing strategies, processes, and operations performance with those of competing organizations

Generic process benchmarking can follow the model of a hotel benchmarking selected processes with a hospital. They share many activities but do not compete with each other. Competitive benchmarking is often facilitated by a trade association or through an independent analyst working on a non-attributed basis. Many smaller organizations conduct more informal benchmarks within their monitoring and measurement program by talking to other organizations at meetings or by hearing the comparisons conducted by mutual customers or groups of stakeholders. Maturity information is easier to benchmark because it does not provide specific process information that could be considered proprietary.

UNCERTAINTY ASSESSMENT

Because there are changes in the internal and external operating environment, it is important to perform internal audits and benchmarks on how well the organization was able to respond to these opportunities and threats once identified. This is likely to involve auditing the uncertainty responses selected after identifying the significant opportunities and threats. Once the uncertainty is addressed, it would be possible to conduct audits on the risk management process focused on determining how effective the organization has been in meeting its strategic objectives.

SELF-ASSESSMENTS

A self-assessment is a "cyclic, comprehensive, systematic and regular review of an organization's activities and results against a model of organizational maturity culminating in planned improvement activities."[4] The cyclic model is like the PDCA cycle as shown in Figure 15.1.

The self-assessments will be performed on each of the elements of the system of management in use by the organization:

- Organizational objectives, worker goals, and action plans
- Scanning of the internal and external operating environment
- Engagement with stakeholders and the social license to operate
- Organizational leadership and governance
- Uncertainty assessment and significant opportunities and threats
- Operational processes and supporting processes
- Managing the system of operations
- Scoping the monitoring and measurement process
- Monitoring and measurement
- Transparency, accountability, and performance
- Self-assessment and maturity
- Continual improvement, innovation, and learning
- Managing organizational resilience

The self-assessment is related to the internal management of the sustainability program as embedded in each of the operational areas listed above. The purpose of conducting the self-assessment is to direct, focus, and motivate various improvement activities and to seek means of innovating and furthering the organization's learning

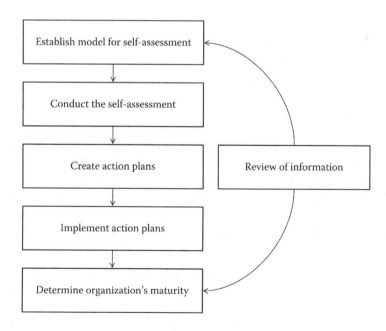

FIGURE 15.1   Self-assessment process.[4]

process, including its knowledge management activity, sense-making, and decision-making. Self-assessment is linked to the operation of the management system and enables the organization to seek interconnections between each of the areas in this system. There are many external reasons to perform a self-assessment (e.g., stakeholder engagement, customer retention, and managing uncertainty). However, many organizations conduct their self-assessments more focused on internal reasons.[4]

Studies have demonstrated that self-assessments provide managers with a valuable method to coordinate and give directions to all the activities that are part of the organization's system of operation.[4] The self-assessments help increase awareness of the importance of effective, efficient, and efficacious strategies. They are seen as a way to improve operational performance.

Overall, self-assessment appears to help organizations manage data gathering, discuss strengths and weaknesses, develop an improvement plan, and link all these activities to document the effectiveness of the sustainability policy and strategy. These activities promote organizational learning by communicating the assessment results and getting feedback from management on the results. The process provides a means for learning the importance of cascading the objectives through the organization and using employee goals and action plans to demonstrate that the objectives are adequately addressed.

Organizations that use self-assessment reported that the most substantial improvements were:[4]

- Improving the reliability of operations and support processes
- Creating a better understanding of the importance and interconnectedness of the elements in the organization's system of management
- Realizing how important it is to maintain and improve the system of management
- Being able to prevent operation losses and realize the value of the significant opportunities from the uncertainty assessment
- Realizing both the cost savings of operational efficiencies and the costs associated with not embedding sustainability into the operations

Self-assessment is a process that leaders use in organizations that is aimed at increasing the effectiveness of processes and the efficiency of processes and creating an efficacious strategy. The self-assessment can be even more effective if combined with the determination of the maturity of the organization's sustainability program.

CREATING A MATURITY MATRIX

Maturity matrices have been used for systems of management in the quality management field for nearly 40 years.[5] The quality maturity matrix was described as a comparison measurement tool used to compare different operations, keeping in mind that the purpose of the comparisons was to focus attention on the areas that were not among the top-performing operations. They are also useful for reporting the comparative results of the program over time.

	Level 1	Level 2	Level 3	Level 4	Level 5
Element 1	Criteria 1 base level			→	Criteria 1 best practice
Element 2	Criteria 2 base level				Criteria 2 best practice
Element 3	Criteria 3 base level				Criteria 3 best practice

FIGURE 15.2 Maturity matrix.[2]

A maturity matrix was presented in an early sustainable development management system standard.[6] The rationale for including the maturity matrix in this standard was to help the organization determine its position along the sustainable development path. It was noted that the maturity matrix is easy to construct and maintain by any type or size organization.

It would make sense that using a maturity matrix provides an effective means for measuring a sustainability program's effectiveness by using the management system supplied in this book. This approach provides the means for measuring the program rather than its drivers or results.

An organization with a mature embedded sustainability program performs its operations effectively and efficiently and achieves sustainability success by:[2]

- Engaging with the internal and external stakeholders
- Monitoring changes in the organization's internal and external context
- Identifying possible scenarios for improvement and innovation
- Defining and deploying policies and strategies through the leaders
- Matching its strategic objectives with operational and tactical objectives
- Constantly improving its supporting process performance with lead indicators
- Managing the processes, resources, and interactions in the suppliers, inputs, process, outputs, and customers (SIPOC) level of the value chain model

Most maturity matrices use five maturity levels (Figure 15.2). The first column contains the program elements of the system of management. The performance criteria are placed in each row to describe the different maturity levels and determine the strengths and weaknesses. The requirements for the level 5 case help the organization understand its need to improve over time. Many organizations use a criteria matrix for a couple of years and then update the criteria to make the level 5 attainment a bit more challenging. This supports continual improvement, innovation, and learning within the organization.

USING THE MATURITY MATRIX

There are several ways to use the maturity matrix for an organizational self-assessment. Some organizations have the leaders conduct the assessment. In other organizations, employees receive training to consistently use this self-assessment method and report the results to management. Larger organizations with a large number of facilities might have the internal audit teams conduct the maturity assessment. If the maturity matrix is independent of the organization's people, operational management and process owners must participate in the evaluation to ensure that the process will capture the organization's behavior and current performance.[7]

The maturity matrix is typically used in a step-by-step process:[7]

1. Define the scope of the self-assessment by deciding what parts of the organization will be included, along with the documentation of the rationale for the selection. The focus of the assessment should also be determined and documented. Here are a few different options:
   a. A self-assessment of the key elements
   b. A self-assessment of the details of the key elements based on the specifications in place (formal or informal) of the organization's system of management
   c. A self-assessment that takes into account different levels of maturity based on a specific standard or benchmark

2. Identify the roles, responsibilities, and authorities of the people involved in the self-assessment and determine when it will be conducted.
3. Determine how the self-assessment will be conducted. For example, who will staff the team that will be used, and will there be a facilitator assigned to provide oversight to the process?
4. Identify the maturity level for each of the organization's processes. This is accomplished by comparing what is observed during the assessment with what is listed in the maturity grids and by marking the elements that the organization is already applying. The current maturity level will be the highest maturity level achieved with no preceding gaps up to that point.
5. Consolidate the results into a report. It should provide a record of progress over time and can be used to present information both internally and externally.
6. Assess the current performance of the organization's processes and operations, identifying the strengths and opportunities for improvement.

An organization can be at different maturity levels for other elements. Reviewing the gaps can help leaders plan and prioritize the improvement or innovation activities required to move individual elements to a higher level. It is essential to see the results of the maturity grid review as the "act" portion of the PDCA cycle. Each report will be used to start the cycle planning option for yet another time to drive continual improvement.

CAPABILITY MATURITY MODEL

Maturity grids should not be confused with capability maturity models. For many people, differentiating between capability maturity models and maturity grids is difficult. It is instructive to look at the key distinctions between the two.[8]

WORK ORIENTATION

Maturity grids apply to organizations and do not specify particular processes. The purpose is to identify any process's characteristics to expect the desired outcome. A capability maturity model identifies the best practices for specific processes and evaluates the maturity of an organization in terms of how many of the practices it has implemented. These models are more of an industrial assessment.

MODE OF ASSESSMENT

An assessment using a maturity grid is structured by using the matrix itself. Levels of maturity are allocated against key aspects of performance or critical activities that are included in the matrix cells. It is easy to see how the performance varies with different program characteristics. A capability maturity model assessment uses questionnaires and checklists to assess an organization's overall performance.

INTENT

Maturity grids tend to be less complex as diagnostic and improvement tools without aspiring to provide certification. Many capability maturity models follow a standard format and are internationally recognized. Therefore, it is easy to use them to certify the organization's performance.

Some business organizations may be using both of these methods.

MATURITY PLOTS

Using the level numbers as scores, it is possible to create radar (spider) plots that are useful in communicating the results. Radar plots provide a helpful way to display the maturity measurements so that they can be visually compared with each other (Figure 15.3). It is possible to have different plot lines for different measurement periods or compare the actual to the planned amounts.

In the self-assessment covered in this chapter, we are measuring the elements of the sustainability program. However, this plot could also be used for measurements of the maturity of the use of the sustainability principles or the scores of the six lead indicators. Various scores can be used at any level in the organization or for the organization as a whole. This does not need to be limited to the maturity measurement.

The maturity measurement represents a better way to express the sustainability program within the stakeholder engagement process. It is challenging for stakeholders to go through several initiative results and figure out just how well everything is going within the company. These maturity examinations and the resultant plots should help detect trends that conditions are improving, or something is backsliding and needs improvement. Quality management programs use these methods to detect changes in the products and services before they lead to market problems. The same should be true with sustainability programs.

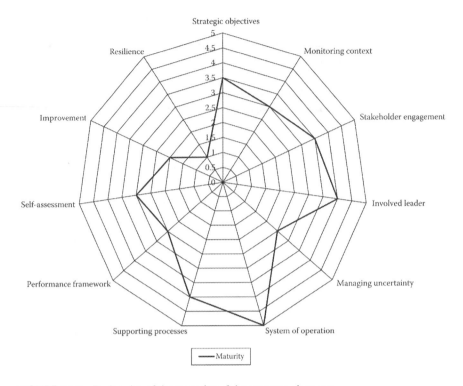

FIGURE 15.3   Radar plot of the maturity of the program elements.

ESSENTIAL QUESTIONS ON SUSTAINABILITY SELF-ASSESSMENT AND MATURITY

1. Why should an organization consider using self-assessment as a means of measuring the performance of its sustainability program?
2. What can other performance tools be used to support the measurement of the performance of an organization's sustainability program?
3. Why is a maturity matrix effective for an organization to continually improve its sustainability program?
4. What are some of the uses for maturity radar (spider) plots for communicating the status of a sustainability program within the stakeholder engagement process?

REFERENCES

1. ISO (International Organization for Standardization), *Guidelines for Auditing Management Systems. ISO 19011*, Geneva, 2011.
2. ISO (International Organization for Standardization), *Risk Management: Principles and Guidelines. ISO 31000*, Geneva, 2009.

3. R. Pojasek, "Benchmarking to sustainability in four steps," *Environmental Quality Management*, vol. 20, no. 2, p. 87–94, 2010.
4. T. B. A. M. R. a. W. D. van der Wiele, "Improvement in organizational performance and self-assessment practices by selected American firms. 7," *Quality Management Journal*, vol. 4, p. 6–22, 2000.
5. P. Crosby, *Quality Is Free: The Art of Making Quality Certain*, New York: Mentor Books, 1979.
6. British Standards, *Guidance for Managing Sustainable Development. BS 8900*, London: British Standards Institute, 2006.
7. ISO (International Organization for Standardization), *Managing for the Sustained Success of an Organization: A Quality Management Approach. ISO 9004*, Geneva, 2009.
8. A. M. J. a. C. P. Maier, "Assessing organizational capabilities," *Reviewing and Guiding the Development of Maturity Grids*, vol. 59, no. 1, p. 138–159, 2012.

16 Improvement, Innovation, and Learning

ABSTRACT

Improvement is essential for an organization to maintain its operational performance and react to changes in its internal and external operating environments. It is a vital part of the "act" component of the plan–do–check–act (PDCA) cycle—taking actions to improve continually.

External change is a driver for innovation. In addition, changes in the organization's external operating environment could require innovation to address the stakeholders' interests. Therefore, an organization must identify the need for innovation and establish an effective process.

Organizational learning includes continual improvement of existing operations and significant change or innovation, often leading to new objectives, products, and services. Through learning, an organization enhances, captures, and formulates its members' knowledge, skills, and experience. This learning can be integrated into what is known as organizational wisdom, shared, and used by the organization to support and enrich its improvement and innovation processes.

Organizations embedding sustainability must continuously develop their capabilities to improve, innovate, and embed these activities into their products, processes, operational structures, way of operating in their context, and management system. The foundation for effective and efficient improvement and innovation processes is learning.

CONTINUAL IMPROVEMENT

An organization defines its strategic objectives to seek improvement of processes, products, and services based on the information it receives from the following activities:[1]

- Monitoring, measurement, and analysis of feedback from its full range of stakeholders
- Monitoring, measurement, and analysis of its operations and supporting processes
- Internal audits and self-assessments
- Suggestions from stakeholders and partners
- Management reviews at the operational and strategic levels

It is also possible for continual improvement to be initiated through collecting data, analyzing information, setting operational and tactical objectives, and implementing corrective actions associated with any nonconformities in the management system.

DOI: 10.1201/9781003255116-16

Continual improvement results from a set of recurring activities that are conducted to enhance the operational performance of an organization.[2] Improvement activities can range from small-step, ongoing improvement to more important breakthrough efforts. An emphasis is placed on continual improvement in an organization through setting worker goals at the tactical level that helps meet the tactical objectives, thus beginning the feedback loop toward meeting the organization's strategic objectives.

The terms "continuous improvement" and "continual improvement" are frequently used interchangeably. However, the American Society for Quality constantly reminds people that there is a difference.[3] "Continual improvement" is a broader term referring to general improvement processes and encompassing many different approaches, covering different areas at different times. Continuous improvement is a subset of continual improvement with a more specific focus on linear, incremental improvement within an existing process. Continual improvement is not limited to quality initiatives. Improvements can be made in organizational strategy, results, and customer, employee, supplier, and stakeholder relationships. It means "getting better all the time." Improvement is a result. It can only be claimed after a beneficial change in an organization's performance.

Improvement can be affected reactively (e.g., corrective action), incrementally (e.g., continual improvement), by step change (breakthrough), or proactively (e.g., innovation).[4] Improvements can be focused on a program (e.g., environmental management), process (e.g., resulting in a product or service), or results. It is common for people to look for areas of underperformance or opportunities to improve the practice of continual improvement and address that. When an improvement project begins, it is possible to track, review, and audit these initiatives' planning, implementation, completion, and results. To address these widespread improvement activities, the quality management system of operation has raised improvement to a principle seeking to create a quality culture that leads to:[1]

- Employing a consistent organization-wide approach to continual improvement
- Making continual improvement of products, services, processes, and systems an objective for every individual in the organization (i.e., embedded in the work)
- Providing people with training in the methods and tools of continual improvement
- Ensuring that people are competent to promote and complete improvement projects successfully
- Establishing goals to guide and measurements to track continual improvement
- Recognizing and acknowledging the improvements that result from these activities

In practical terms, some improvements take time to achieve. Some may require allocations for a budget or careful planning for a consistent rollout of the results.

Therefore, improvement plans should consider priorities and relative benefits and allow for progress monitoring.[5] Continual improvement of the organization's overall performance should be a strategic objective of the organization.[1] It is widely believed that organizations with an ongoing focus on improvement are successful in many ways such success is measured.

The leaders of the organization need to ensure that continual improvement becomes established as part of the organization's culture through several directed efforts by:[1]

- Providing opportunities for the internal stakeholders to embed improvement into their work
- Providing the necessary resources for this to happen
- Recognizing and rewarding successes
- Continually seeking to improve the effectiveness and efficiency of the improvement process

Breakthrough projects can be conducted using a combination of project management methods and approaches for managing change while seeking to introduce these methods into the organization's culture.

It might help to look at the basis for determining the maturity of an organization's innovation process:[1]

- Level 1: Improvement activities are ad hoc and based on customer or regulatory complaints.
- Level 2: Basic improvement processes based on corrective and preventive actions are in place. The organization provides awareness for its members to help them understand the concept and practice of continual improvement.
- Level 3: Improvement efforts can be demonstrated in most of the organization's processes, products, and services. The focus of the improvement process is aligned with the strategy and the strategic objectives. Recognition systems are in place for members to generate strategically relevant improvements in their work. Continual improvement processes work at some levels of the organization and with its suppliers and partners.
- Level 4: Results generated from the improvement processes enhance the organization's performance. The improvement processes are systematically reviewed. Improvement is applied to processes, products, services, organizational structures, the operating model, and the organization's management and supporting processes system.
- Level 5: There is evidence of a strong relationship between improvement activities and the achievement of above-average performance for the organization determined in benchmarking activity. Improvement is embedded as a routing activity across the entire organization and its suppliers and partners. The focus is on improving the organization's performance, including its ability to learn and change.

INNOVATION

Changes in the organization's internal and external operating environments may require innovation to address the stakeholders' interests and continually improve its ability to meet its strategic objectives.[1] Leaders should identify these changes and initiate the impetus for innovation by refining the strategic plan so that it continues its efficacious ways.

Innovation can be applied within the organization through changes in:[1]

- Technology, product, or service to not only respond to the changing needs and interests of stakeholders but also to anticipate potential changes in the organization's external operating environment and product and service life cycles
- Processes that improve cycle time, stability, and variation
- The organization itself, through innovation in organizational structure and governance
- The organization's system of management that ensures the competitive advantage is maintained and new opportunities are realized when there are emerging changes in the organization's context

Planning for innovation should drive the organization to check its ability to afford the proposed activity by prioritizing the information and resources needed for this purpose. The resources that the organization should consider include:[1]

- Creative ideas from all the engaged stakeholders
- Results from management reviews of its strategy
- Results of activities to improve its system of management and supporting processes
- Organization's performance
- Results of assessments commissioned by the leaders
- Evaluating the threats, as well as the opportunities for innovation
- Internal skills and knowledge, as well as assistance from suppliers, partners, customers, and stakeholders
- Availability of scientific or technical information
- Products and services that are approaching the end of their productive time and the need for a replacement product and service offerings
- Availability of methods for innovation while considering both the opportunities and threats

The design, implementation, and management of processes for innovation can be influenced by:[1]

- The urgency of the need for innovation
- Innovation objectives and their impacts on processes, products, services, and the organization's structure

- Understanding of the organization's current situation and current capabilities in relation to its innovation objectives
- The leader's commitment to innovation
- Willingness to challenge the status quo
- Availability and emergence of new technologies
- Identification of the threats associated with innovation
- Successful exchange of knowledge and expertise internally and outside the organization

The timing for introducing an innovation process is usually a balance between the urgency with which it is needed and the resources available for effecting change.

Innovation based on the organization's learning ability is essential for the long-term success of its sustainability program. The organization should use a process that aligns with its sustainability strategy to prioritize innovations in conjunction with its uncertainty assessment process. Innovation must be supported with the resources necessary or appropriate. Innovation has become imperative for any organization operating in the changing world. Since almost all organizations are dealing with frequent and significant changes, they need a defined process for developing and maintaining innovation across everything they do daily.

It might help to look at the basis for determining the maturity of an organization's innovation process:[1]

- Level 1: There is limited innovation. New products and services are introduced on an ad hoc basis with no planning for innovation.
- Level 2: Innovation activities are based on data concerning the needs and interests of customers and other stakeholders.
- Level 3: The innovation process for new processes, products, and services can identify changes in the organization's external operating environment to initiate the planning for innovation.
- Level 4: Innovations are prioritized using uncertainty assessment based on the balance between their urgency, the availability of resources, and the organization's strategy. Suppliers and partners are involved in the innovation processes. The innovation processes' effectiveness and efficiency are regularly assessed as part of the learning process (see the next section, "Organizational Learning"). Innovation is used to improve the way the organization operates.
- Level 5: Innovation activities anticipate possible changes in the external operating environment. Preventive plans are developed to avoid or minimize the identified opportunities and threats associated with the innovation activities. Innovation is applied to processes, products, services, organizational structures, the operating model, and the organization's system of management, as well as its supporting processes.

Many consider embedding sustainability into the organization an essential part of the innovation process. The new work items for the organization may include:[6]

- Facilitate reflection on the consideration of sustainability interests in the innovation process
- Assess the level of embedding of these interests in the innovation process
- Identify points of vigilance and critical points to facilitate the embedding of interests at each stage of the process
- Propose an evaluation system that is consistent with sustainability interests

Innovation is essential in today's world for all organizations. This globalized context implies that implementing partnerships and sharing a common vocabulary and practices have become critical for sustainable development.

ORGANIZATIONAL LEARNING

The organization should encourage improvement and innovation through learning. In this way, the organization can enhance, capture, and formulate its members' knowledge, skills, and experience and integrate them into organizational wisdom, which can then be shared and used by the organization to foster its improvement and innovation processes.[1]

Development of the organization's learning ability depends on its ability to collect information, analyze, and gain insights from various activities in its internal and external operating environments, as well as its ability to integrate the knowledge and to think of its members into the value system of the organization.

To be successful, organizational learning is embedded in the way an organization operates, just like sustainability. This means that learning is:[7]

- Part of what members and employees do every day
- Practiced at personal, work unit, and organizational levels
- About solving problems at their source (i.e., root cause analysis)
- Focused on building and sharing knowledge and participation in sense-making
- Driven by opportunities to create significant change and innovate

Learning is achieved through research, evaluation, improvement cycles, ideas from employees and external stakeholders, sharing best practices with partner organizations, and benchmarking. Learning contributes to the organization having a competitive advantage and sustainability.

Organizational learning can result in the following:[7]

- Enhanced value to customers through new and improved products and services
- Development of new organizational and business opportunities
- Development of new, improved, or disruptive processes or business models
- Reduced errors, defects, wastes, and their related costs
- Improved process responsiveness and the cycle time required for products and services

- Increases in resource productivity resulting in requirements for fewer resources as inputs
- Enhanced performance in fulfilling the organization's sustainability policy

Learning that integrates the capabilities of the organization and its members can be achieved by combining knowledge, thinking patterns, and behavior patterns of people with the organization's values.[1] This involves the consideration of:[1]

- The organization's mission, vision, values, and strategy
- Supporting efforts in learning and demonstrating the leaders' involvement
- Stimulations of networking, connectivity, interactivity, and sharing of knowledge both inside and outside the organization
- Maintaining systems for learning and sharing of knowledge
- Recognizing, supporting, and rewarding the improvement of people's competence through processes for learning and sharing knowledge
- Appreciation of creativity, supporting the diversity of opinions of the different people in the organization

Rapid access to and use of this knowledge can enhance the organization's ability to manage and maintain its sustainability program.

It might help to look at the basis for determining the maturity of an organization's innovation process:[1]

- Level 1: Some lessons are learned due to complaints and the difficulty of maintaining the social license to operate. Learning is on an individual basis without the sharing of knowledge.
- Level 2: Learning is generated reactively from systematically analyzing problems and other data. Processes exist for the sharing of information and knowledge.
- Level 3: There are planned activities, events, and forums for sharing information. A system is in place for recognizing positive results from suggestions or lessons learned. Learning is addressed in the strategy and policies.
- Level 4: Learning is a key process. Networking, connectivity, and leaders stimulate interactivity to share knowledge in the organization. Leaders support initiatives for learning and lead the process by example. The organization's learning ability integrates personal competence and organizational competence. Learning is fundamental to the improvement and innovation processes.
- Level 5: The culture of learning permits the taking of risks and the acceptance of failure, provided this leads to learning from the mistakes, threats, and opportunities that are identified. There are external engagements for learning.

ESSENTIAL QUESTIONS ON IMPROVEMENT, INNOVATION, AND LEARNING

1. Why is it important for an organization to seek continual improvement of its processes, products, and services?
2. What is the key driving force that creates the impetus for an organization to improve its innovation process?
3. What is the relationship between improvement, innovation, and learning, and why is this relationship important to the development of effective learning?
4. How does a maturity matrix help an organization perform a self-assessment and measure its progress in improvement, innovation, and learning?

REFERENCES

1. ISO (International Organization for Standardization), *Managing for the Sustained Success of an Organization: A Quality Management Approach. ISO 9004*, Geneva, 2009.
2. Praxiom, *Plain English Definition ISO 9000:2015*, Calgary: Praxiom Research Group Ltd., 2014. [Online]. Available: http://www.praxiom.com/iso-definitions.htm.
3. ASQ (American Society for Quality), "Basic concepts: Continuous improvement," ASQ (American Society for Quality), n.d. [Online]. Available: http://www.asq.org/learn-about-quality/continuous-improvement/overview/overview.html.
4. Geneva, *Quality Management Systems: Requirements. ISO 9001*, Geneva: ISO (International Organization for Standardization), 2015.
5. ISO (International Organization for Standardization), *Risk Management: Guidance for the Implementation of ISO 31000. ISO/TR 31004*, Geneva, 2013.
6. ISO (International Organization for Standardization), *Innovation Process: Interaction, Tools and Methods. ISO TS/P233*, Geneva, 2012.
7. NIST/BPEP, "Baldrige (Baldrige performance excellence program)," Gaithersburg: National Institute for Science and Technology, 2013.

17 Organizational Resilience

ABSTRACT

When studying sustainability, one learns about resilient ecosystems, resilient infrastructure, resilient individuals, resilient communities, and resilient value chains. In addition, there has been a growing body of knowledge on resilient organizations. Every day, the sustainability of an organization is tested in a world that constantly changes. Strategic objectives that help an organization survive and prosper contribute to organizational resilience. Highly resilient organizations are more adaptive, competitive, agile, and robust than less resilient organizations. Resilient organizations anticipate, prepare for, respond to, and adapt to disruptive events and other effects of uncertainty. Resilience operates within a risk landscape, conditioned by people, knowledge, technology, and interactions. Resilience is enhanced by integrating and coordinating the various operational programs commonly found in a sustainability program. As a result, resilience is built within the organization and across its value chains and the more extensive web of interactions with other organizations.

DIMENSIONS OF ORGANIZATIONAL RESILIENCE

The study of resilient organizations suggests that this resilience is a function of three interdependent attributes:[1]

- Leadership and culture: The adaptive capacity of the organization
- Networks: The ability to leverage the internal and external relationships that have been developed
- Change readiness: The results of planning and direction that helped establish the ability of the organization to be change-ready

The core strategic purpose of resilience is to enable an organization to prepare itself to survive and prosper. These purposes include the following:[2]

- Competitiveness: Being able to continue past, recover, learn from, and even capitalize on opportunities presented by disruptions in a way that increases the value that exceeds that of the less resilient competitors.
- Coherence: Align strategic objectives with operational resilience measures, such as aligning organizational silos to become more integrated and interoperable.
- Efficiency and effectiveness: Creating a framework to mesh together diverse components while allocating resources to improve overall resilience.

DOI: 10.1201/9781003255116-17

- Reputation: A coherent framework helps build trust among the various operating units and outside allies, helping manage and enhance the organization's reputation.
- Community resilience: Organizational resilience helps to enhance community resilience by assuring stakeholders (e.g., regulators, government, customers, partners, and families) of its ability to provide vital products and services to the public in times of need.

Every organization is subject to the effects of uncertainty, and resilience thinking must embrace learning. Achieving favorable outcomes in the face of uncertainty requires creativity and innovation. An organization must be able to overcome strategic barriers to answer three basic questions after experiencing a disruptive event:[3]

1 What have we learned?
2 How did we learn?
3 How can we integrate what we learned to understand complex, interrelated events; engage in reflective conversations; and cultivate shared aspirations for the future?

As organizations build on the past to expand their ability to act, they are developing new ranges of possible actions that will enable them to meet threats posed by disruptive events. They are even looking for opportunities, often masked by the threats that demand immediate action. As a result, responding without hesitancy to unexpected events is now considered an organizational imperative.[3]

ORGANIZATIONAL FOUNDATIONS FOR RESILIENCE

To build resilience into the operating structure of an organization, it is necessary to create a foundation that helps define the attitudes that affect organizational decisions and actions.[2] This foundation should support the further development of resilience in organizations.

Organizations must be able to detect, assess, prevent, respond to, and recover from disruptive events and challenges of all types. However, organizational resilience differs significantly from the more traditional business continuity concepts. Resilience seeks to create critical thinking, learning, and capabilities not just to bounce back (as with the practice of continuity) but also to bounce forward. This requires a combination of continuity and adaptability. The concept of adaptability includes a strong sense of societal security where pre-established relationships, learning, flexibility, and a sense of new normality are built into the organizations so they can participate in this new focus on continuity and adaptability.[2] After all, the organization is the basic building block of society.

It would be instructive to take the framework for sustainability and see how it can be adjusted and adapted to include this continuity–adaptability perspective.

RESILIENCE IN THE STRUCTURAL OPERATING FRAMEWORK

The organization needs to have a shared vision and purpose (included in the mission statement) for the future (included in the sustainability program). This enables the organization to build resilience into the strategic objectives so that challenge, change, and opportunity are assessed against the vision and mission and can be acted upon accordingly.[2]

Leaders of the organization need to understand how the organization's efficacious strategy as created for the sustainability program can address the needs of resilience development. This understanding starts with knowledge of the current level of resilience.

The organization must promote and create universally shared expectations that strengthen its resilience. This may be reflected in one of the strategic objectives. It can be accomplished by developing and promoting cultural norms that value openness in reviewing and evaluating the organization's resilience and how resilience will be addressed in the feedback loop to ensure that the strategic objectives are met in an uncertain world.

CONTEXT OF THE ORGANIZATION

For resilience to be effective, the organization must be highly informed about its internal and external operating environments. As a result, there is a new focus on influences, factors, opportunities, and threats that might influence or compromise the organization's resilience. Sometimes, this is referred to as situational awareness in the resilience literature.

In addition to the scanning performed as part of the sustainability program, the organization needs to pay attention to methods designed to identify opportunities and threats that fall within the resilience realm:[2]

- Identifying what values and purpose it wants to protect and any threats that could impact those sets of values and purposes
- Looking for emerging factors, along with the other scanning targets, to find new opportunities and threats affecting resilience
- Drawing upon sense-making with the assistance of the stakeholder engagement process and the knowledge management function that is constantly examining opportunities and threats both from the context and from the literature
- Using uncertainty assessment and response to help use the opportunity response to offset the threats before they contribute to the level of uncertainty that lies between the organization and its ability to meet its strategic objectives

It is also essential to scan for the kinds of events that might be expected to affect the organization. An event is defined as an occurrence or change of a particular set of circumstances and their effect on the organization. That event could have one or more occurrences and several causes. An event could also consist of something not happening. Some events do not have consequences and are referred to as a "near miss," "incident," or "close call." Environmental management and health and safety management have preparedness and emergency response programs to deal with events. There are also management programs for business continuity. All these programs will be included in the organization's response to resilience.

STAKEHOLDERS AND THE SOCIAL LICENSE TO OPERATE

The stakeholders are also dealing with uncertainty. They understand how it can affect their organizations. However, they may not understand the power of resilience to help deal with uncertainty. Sharing information on resilience within the stakeholder engagement process will truly be valued. The ability of the stakeholders to assist with a lookout for a change in the external operating environment and sense-making will provide the organization with their appreciation for advice on using the adaptive capacity of resilience. By being transparent on the topic of resilience within the stakeholder engagement process, the organization can help build greater trust. This should help the organization secure and maintain its social license to operate from the stakeholders.

LEADERSHIP AND COMMITMENT

Leaders need to consider the impact of their decisions and the sustainability strategy on an ongoing basis. They need to create a culture in which it is essential to consider resilience within the decision-making process. This culture of trust, openness, and innovation will empower the members or employees in an organization to assume ownership of and address uncertainty as scans of the external operating environment note new opportunities and threats. The leaders delegate authority and responsibility to the individuals best able to decide for the organization, both under normal operations and in crisis.

The leaders are responsible for leading the engagement with all stakeholders. The engagement process needs to foster transparency that enables information to be proactively shared across internal boundaries with independent partners and key stakeholders.

Effective governance enables the organization to exploit the uncertainty assessment and response results. This should direct the internal stakeholders to make decisions following the knowledge and sense-making that has been increasingly involved in assembling knowledge on resilience. Effective governance also enables the organization to encourage improvement and innovation to help continually improve the knowledge system. Resilience will succeed when the governance is coherent, transparent, and "forward-looking" and is embedded in a culture that is supportive of the continual enhancement of organizational resilience.

The leaders are accountable for ensuring that the organization achieves an appropriate level of resilience. This accountability will begin with a resilience policy embedded in the sustainability policy and coordinated with other important policies.

This provides the leaders' commitment to all the outcomes important to the organization, including sustainability and resilience.

MANAGING UNCERTAINTY AND ORGANIZATIONAL PLANNING

The organization must adapt to changing conditions as they emerge in the internal and external operating environments. The organization may accept that some disruptions cannot be prevented and will still occur. It can plan, implement, test, and review a range of measures that will help the organization prepare to deal with disruption, possibly arising from unforeseen events, or effectively adapt when the established plan does not cover what is being experienced.[2]

The people responsible for uncertainty assessment must have open communications with those responsible for operations, supporting processes, and innovation, along with others who need to respond to the changes. Leadership must be involved in this effort to ensure the ability to take timely and informed actions to intercept and contain adverse events, both foreseen and unforeseen, such as to effectively respond to the uncertainty, including overwhelming crises that threaten the continued existence of the organization.[2]

ORGANIZATIONAL SYSTEM OF OPERATING

Perhaps the most important element of resilience is the need to develop coherence in the organization's operations that provide its products and services. All its activities and processes work well together when an organization has coherence. Even simple organizations have many processes to manage. Management systems now have a harmonized structure that makes it easy to add various operational components in a freestyle, mix-and-match manner. On the other hand, it is also easy for an organization to remove operational elements that are no longer being used. Leaders must align operational activities within this structure to achieve coherence and build resilience. To ensure that organizational silos support resilience, the organization needs to embed sustainability and risk management activities and operational disciplines to drive integration. Knowledge also needs to be actively shared across internal organizational boundaries so that opportunities and threats are addressed coherently by all parts of the organization.

Sustainability and resilience draw on an extensive list of different processes. Unfortunately, for many organizations, these processes are operated in silos. Leaders should find a way to create bridges between these operating silos using the process approach.

Understanding the interdependencies with other organizations, including suppliers, contractors, outsourcers, and competitors, is essential.[2] There is a greater need for interdependencies when developing resilience methods.

The organization needs to adapt its operations to changing conditions as they emerge in the internal and external operating environments. This may involve switching between preplanned responses to events and adaptive actions as necessary and modifying its governance structures, operations, and behaviors to adjust to new conditions.[2] The planning for the operations focuses on typical situations. There must be a link between the uncertainty assessment and response efforts in the planning and emergency planning for abnormal events in the operations. This will allow the organization to respond to change in a resourceful manner as the support processes help operations adjust to the new conditions.[2]

To strengthen the organization's operations, specific measures need to be implemented to address disruptive events, emergent risks, and changes in the external operating environment. This can be addressed by ensuring that the employees understand the importance of considering resilience during decision-making and other instances of change management. In addition, all employees may need to take actions to prevent or reduce the likelihood of disruptive events or disruption to protect people, physical assets, financial value, reputation, and social capital.

ORGANIZATIONAL SUPPORTING OPERATIONS

Organizations need people competent in resilience and adaptability through education, training, and experience to develop and implement uncertainty assessment and business continuity plans. Employees need to identify significant threats and opportunities associated with their work and apply procedures to reduce the consequences of a disruptive event. Because many organizations will not experience major disruptions, experience can be achieved through exercise and rehearsed drills. The exercises need to be modified regularly to consider new information in the knowledge management system. However, organizations and their members must be able and willing to adapt to change to become resilient. When a significant disturbance strikes, continuity plans may need to be radically adjusted to reflect the new circumstances. In some cases, the business continuity plans will need to be discarded to ensure appropriate and considered action is taken.[4]

The organization needs to create the means, incentives, and imperatives to communicate how resilience is a unifying factor in the organization's management system. These messages will create a need for engaged employees and outside stakeholders who want to know more. Changes associated with resilience will lead to collaboration between those involved in these stakeholder engagements, collaboration across the value chain, and even among competitors.

The importance of building adaptive capacity must be included in the awareness program that is operated by the supporting operations. They need to disseminate and oversee the implementation of good practices for dealing with resilience identified within and outside the organization. It is useful to have an effort to share information on errors, failures, and mistakes transparently so that all the internal stakeholders are aware and learn how to avoid these issues. Many organizations share information through trade and professional associations to gain valuable lessons on dealing with uncertainty and improving operations. This activity helps all organizations improve. Finally, the adaptive capacity can be improved by developing the internal stakeholders to get involved in the innovation process and to know that there is a need to be flexible during times of change.[2]

RESILIENCE IN ORGANIZATIONAL PERFORMANCE MANAGEMENT

The organization needs to include efforts to address resilience when it audits its system of operation to demonstrate the program's efficacy. The resilience measures need to be tied to the capacity of people to learn and adapt when required. This information should inform the organization's understanding of its resilience capability and may be used to prompt strategic change to enhance how resilience drives strategic change. As part of these efforts, the organization should verify its compliance with legal and regulatory obligations.

MONITORING AND MEASUREMENT OF PERFORMANCE

The organization must define the transparency and levels of accountability by which individual and collective decisions and actions on resilience are related to the norms, expectations, and obligations of the organization, its partners, and its stakeholders. In addition, there needs to be a means of measuring the performance of the resilience efforts within the framework of measuring the standard operations, products, and services.

There should be a self-assessment and maturity matrix for resilience. This would enable the organization's leaders to determine the level of resilience already in place and how new efforts to enhance resilience are leading to an increase in resilience maturity. To affect this assessment, the organization should identify:[2]

- What needs to be monitored and measured
- The methods for monitoring, measurement, analysis, and evaluation, as applicable, to ensure valid results
- How to provide a continuous assessment of resilience
- The thresholds at which the output from measurements will be considered acceptable
- How measurement and monitoring arrangements will work alongside, support, or integrate into existing monitoring processes
- How the results from monitoring and measurement will be analyzed and evaluated

The organization needs to understand what evidence it requires to support its resilience assessment and ensure an evaluation process is developed to support the evidence.

IMPROVEMENT, INNOVATION, AND LEARNING

There needs to be a continual improvement of the organization's efforts to address resilience in combination with the sustainability program.

Innovation must be tied to the organization's adaptive capacity in its operations. During uncertainty assessment and response, there needs to be a direct link with the people responsible for the innovation (e.g., introducing new materials, ideas, or products and services). This will enable the organization to exploit the opportunities and offset the unexpected threats during these scanning or sense-making activities. This may lead to new or improved products and services that fit the new conditions brought about by long-term changes in the operating environment. In this way, innovation can serve the organization by identifying new and better solutions and addressing the organization's shifting needs arising from its ever-changing external operating environment.[2]

Learning is critical in the practice of resilience. An organization should constantly improve its knowledge and sense-making to incorporate resilience in decision-making.

ESSENTIAL QUESTIONS ON RESILIENT ORGANIZATIONS

1. How can an organization introduce resilience into its strategic objectives to help it cope with uncertainty and unanticipated disruptions over the long term?
2. How can the organization build the capacity to adapt to changing conditions as they emerge?
3. Why is it essential for the organization to bring coherence to its system of operations and supporting processes?
4. Why should an organization embed resilience into its system of operations rather than create a new program to address disruptions to its operations?

REFERENCES

1. Erica Seville, "Resilient organizations: How to Survive, Thrive and Create Opportunities Through Crisis and Change,"Kogan Page, 2016
2. British Standards Institute, *Guidance on Organizational Resilience BS 65000*, London, 2014.
3. O. Serrat, "On resilient organizations," *Knowledge Solutions*, Vols. Retrieved June 29, 2015. Manila: Asian Development Bank, 2013.
4. A. McAsian, "Organisational resilience: Understanding the concepts and its application," Adelaide: Torrens Resilience Institute, 2010.

Index